An Introduction to Zoology

An Introduction to Zoology

Professor Manoranjan Prasad Sinha

SYRAWOOD
PUBLISHING HOUSE
New York

Published by Syrawood Publishing House,
750 Third Avenue, 9th Floor,
New York, NY 10017, USA
www.syrawoodpublishinghouse.com

An Introduction to Zoology
Professor Manoranjan Prasad Sinha

International Standard Book Number: 978-1-68286-784-6 (Hardback)

Cataloging-in-Publication Data

An Introduction to Zoology / Professor Manoranjan Prasad Sinha.
p. cm.
Includes bibliographical references and index.
ISBN 978-1-68286-784-6
1. Zoology. 2. Animals. 3. Biology. I. Sinha, Professor Manoranjan Prasad.
QL45.2 .I58 2019
590--dc23

Printed in China.

TABLE OF CONTENTS

PREFACE

The branch of biology, which encompasses a comprehensive study of the animal kingdom and explores the structure, classification, embryology, evolution and distribution of animals, is referred to as zoology. In the Linnaean classification system, animals are grouped into the classes of Mammalia, Amphibia, Insecta, Pisces, Aves and Vermes. Some of the major sub-disciplines of zoology are animal physiology, invertebrate zoology, vertebrate zoology, behavioral ecology, etc. This textbook provides comprehensive insights into the field of zoology. Some of the diverse topics covered in this book address the varied branches that fall under this category. It is a complete source of knowledge on the present status of this important field.

A detailed account of the significant topics covered in this book is provided below:

Chapter 1- Animals are the multicellular eukaryotic organisms that constitute the biological kingdom of Animalia. The study of animals and their characteristics is under the domain of zoology. This is an introductory chapter, which will introduce briefly all the significant aspects of zoology.

Chapter 2- Animals that have a backbone, such as reptiles, amphibians, fish, mammals and birds, are called vertebrates. They are studied under the domain of vertebrate zoology. Mammalogy, ornithology, herpetology and ichthyology are some of the significant subdisciplines of vertebrate zoology, which have been elaborated in this chapter.

Chapter 3- Invertebrate zoology is a subdiscipline of zoology, which studies animals that do not have a backbone. Arthropodology, Cnidariology, Helminthology and Malacology are some of the subdivisions of invertebrate zoology. They have been elaborated in this chapter to develop a better perspective about the branches of invertebrate zoology.

Chapter 4- Animal behavior is studied under the domain of ethology. In order to completely understand animal behavior, it is necessary to understand the processes related to it. The following chapter elucidates the topics such as animal communication, consciousness, emotions, etc.

It gives me an immense pleasure to thank our entire team for their efforts. Finally in the end, I would like to thank my family and colleagues who have been a great source of inspiration and support.

Professor Manoranjan Prasad Sinha

Chapter 1

Understanding Zoology

Animals are the multicellular eukaryotic organisms that constitute the biological kingdom of Animalia. The study of animals and their characteristics is under the domain of zoology. This is an introductory chapter, which will introduce briefly all the significant aspects of zoology.

Zoology (also known as animal science) is the branch of biology devoted to the study of animal life. It covers areas ranging from the structure of organisms to the subcellular unit of life. Some zoologists are interested in the biology of particular groups of animals. Others are concerned with the structure and function of animal bodies. Still others study how new animals are formed and how their characteristics are passed on from one generation to another. Zoologists study the interactions of animals with one another and their environments, as well as the significance of the behavior of animals.

Zoology is both descriptive and analytical. It can be approached either as a basic science or as an applied science. A worker in basic zoology is interested in knowledge of animals for its own sake without consideration of the direct application of the information gained. In contrast, workers in applied zoology are interested in information that will directly benefit humans and animals (medicine, for example).

Historically, the study of zoology can be viewed as a series of efforts to analyze and classify animals. The ancient Greek philosopher Aristotle is credited with devising the system of classifying animals that recognized similarities among diverse organisms in the fourth century B.C.E.; he arranged groups of animals according to mode of reproduction and habitat. Zoology began to emerge as a science in the 12th century and long was dominated by studies of anatomy and efforts at classifying animals. The Swedish botanist Carolus Linnaeus developed a system of nomenclature that still is used today — the binomial system of genus and species — and established as a discipline taxonomy, the science of classification according to a predetermined system.

Zoology today is as diverse as the animal kingdom it studies, broadening its range to include such fields as genetics and biochemistry. It now is considered an interdisciplinary field that applies a great variety of techniques to obtain knowledge of the animal kingdom. For instance, the genetic study of DNA from various animals can provide insights into their evolutionary history. Zoologists who concentrate on the morphology (the study of structure, including muscles, bones, cells and cellular components) employ many techniques first developed in the biochemistry lab.

Research

Structural

Cell biology studies the structural and physiological properties of cells, including their behavior, interactions, and environment. This is done on both the microscopic and molecular levels, for single-celled organisms such as bacteria as well as the specialized cells in multicellular organisms such as humans. Understanding the structure and function of cells is fundamental to all of the biological sciences. The similarities and differences between cell types are particularly relevant to molecular biology.

Anatomy considers the forms of macroscopic structures such as organs and organ systems. It focuses on how organs and organ systems work together in the bodies of humans and animals, in addition to how they work independently. Anatomy and cell biology are two studies that are closely related, and can be categorized under "structural" studies.

Physiological

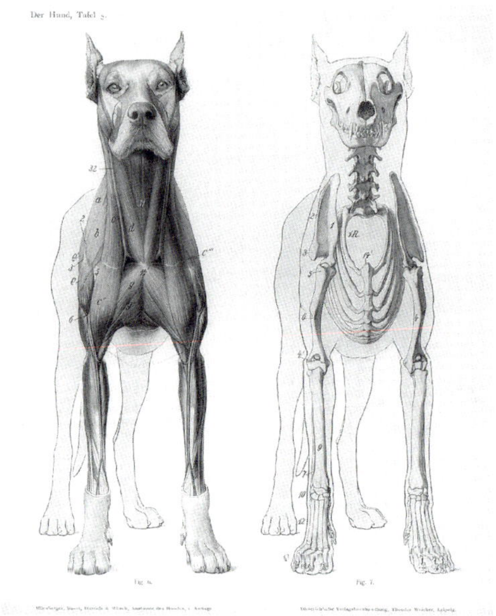

Animal anatomical

Physiology studies the mechanical, physical, and biochemical processes of living organisms by attempting to understand how all of the structures function as a whole. The theme of "structure to function" is central to biology. Physiological studies have traditionally been divided into plant physiology and animal physiology, but some principles of physiology are universal, no matter what particular organism is being studied. For example, what is learned about the physiology of yeast cells can also apply to human cells. The field of animal physiology extends the tools and methods of human physiology to non-human species. Physiology studies how for example nervous, immune, endocrine, respiratory, and circulatory systems, function and interact.

Evolutionary

Evolutionary research is concerned with the origin and descent of species, as well as their change over time, and includes scientists from many taxonomically oriented disciplines. For example, it generally involves scientists who have special training in particular organisms such as mammalogy, ornithology, herpetology, or entomology, but use those organisms as systems to answer general questions about evolution.

Evolutionary biology is partly based on paleontology, which uses the fossil record to answer questions about the mode and tempo of evolution, and partly on the developments in areas such as population genetics and evolutionary theory. Following the development of DNA fingerprinting techniques in the late 20th century, the application of these techniques in zoology has increased the understanding of animal populations. In the 1980s, developmental biology re-entered evolutionary biology from its initial exclusion from the modern synthesis through the study of evolutionary developmental biology. Related fields often considered part of evolutionary biology are phylogenetics, systematics, and taxonomy.

Classification

Scientific classification in zoology, is a method by which zoologists group and categorize organisms by biological type, such as genus or species. Biological classification is a form of scientific taxonomy. Modern biological classification has its root in the work of Carl Linnaeus, who grouped species according to shared physical characteristics. These groupings have since been revised to improve consistency with the Darwinian principle of common descent. Molecular phylogenetics, which uses DNA sequences as data, has driven many recent revisions and is likely to continue to do so. Biological classification belongs to the science of zoological systematics.

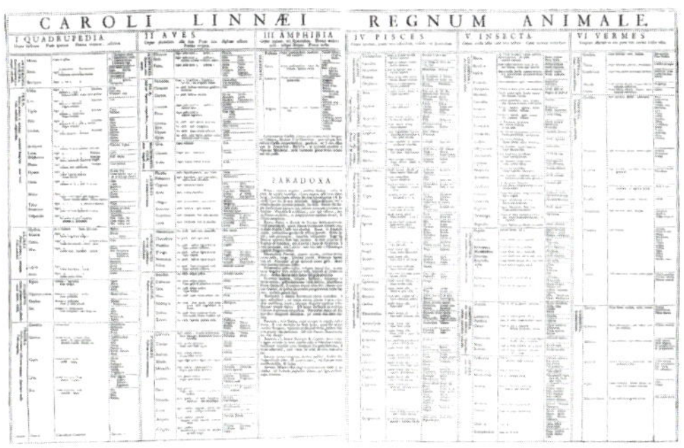
CAROLI LINNÆI REGNUM ANIMALE.
I QUADRUPEDIA | II AVES | III AMPHIBIA | IV PISCES | V INSECTA | VI VERMES

Linnaeus's table of the animal kingdom from the first edition of *Systema Naturae*.

Many scientists now consider the five-kingdom system outdated. Modern alternative classification systems generally start with the three-domain system: Archaea (originally Archaebacteria); Bacteria (originally Eubacteria); Eukaryota (including protists, fungi, plants, and animals) These domains reflect whether the cells have nuclei or not, as well as differences in the chemical composition of the cell exteriors.

Further, each kingdom is broken down recursively until each species is separately classified. The order is: Domain; kingdom; phylum; class; order; family; genus; species. The scientific name of an organism is generated from its genus and species. For example, humans are listed as *Homo sapiens*. *Homo* is the genus, and *sapiens* the specific epithet, both of them combined make up the species name. When writing the scientific name of an organism, it is proper to capitalize the first letter in the genus and put all of the specific epithet in lowercase. Additionally, the entire term may be italicized or underlined.

The dominant classification system is called the Linnaean taxonomy. It includes ranks and binomial nomenclature. The classification, taxonomy, and nomenclature of zoological organisms is administered by the International Code of Zoological Nomenclature. A merging draft, BioCode, was published in 1997 in an attempt to standardize nomenclature, but has yet to be formally adopted.

Ethology

Kelp gull chicks peck at red spot on mother's beak to stimulate the regurgitating reflex.

Ethology is the scientific and objective study of animal behavior under natural conditions, as opposed to behaviourism, which focuses on behavioral response studies in a laboratory setting. Ethologists have been particularly concerned with the evolution of behavior and the understanding of behavior in terms of the theory of natural selection. In one sense, the first modern ethologist was Charles Darwin, whose book, *The Expression of the Emotions in Man and Animals,* influenced many future ethologists.

Biogeography

Biogeography studies the spatial distribution of organisms on the Earth, focusing on topics like plate tectonics, climate change, dispersal and migration, and cladistics. The creation of this study is widely accredited to Alfred Russel Wallace, a British biologist who had some of his work jointly published with Charles Darwin.

Branches of Zoology

Although the study of animal life is ancient, its scientific incarnation is relatively modern. This mirrors the transition from natural history to biology at the start of the 19th century. Since Hunter and Cuvier, comparative anatomical study has been associated with morphography, shaping the modern areas of zoological investigation: anatomy, physiology, histology, embryology, teratology and ethology. Modern zoology first arose in German and British universities. In Britain, Thomas Henry Huxley was a prominent figure. His ideas were centered on the morphology of animals. Many consider him the greatest comparative anatomist of the latter half of the 19th century. Similar to Hunter, his courses were composed of lectures and laboratory practical classes in contrast to the previous format of lectures only.

Gradually zoology expanded beyond Huxley's comparative anatomy to include the following sub-disciplines:

- Zoography, also known as *descriptive zoology*, is the applied science of describing animals and their habitats
- Comparative anatomy studies the structure of animals
- Animal physiology
- Behavioral ecology
- Ethology studies animal behavior
- Invertebrate zoology
- Vertebrate zoology
- Soil zoology
- The various taxonomically oriented disciplines such as mammalogy, biological anthropology, herpetology, ornithology, and entomology identify and classify species and study the structures and mechanisms specific to those groups.

Related fields:

- Evolutionary biology: Development of both animals and plants is considered in the articles on evolution, population genetics, heredity, variation, Mendelism, and reproduction.
- Molecular biology studies the common genetic and developmental mechanisms of animals and plants
- Palaeontology
- Systematics, cladistics, phylogenetics, phylogeography, biogeography, and taxonomy classify and group species via common descent and regional associations

Animal

Animals are creatures in the kingdom Animalia, one of the five major divisions of organisms. They are multicelled, eukaryotic organisms, meaning their cells contain nuclei and other structures called organelles, all of which are enclosed by thin membranes. (Eukaryote means "true nucleus.") Unlike plants, their cells do not have cell walls. Animals are capable of moving their bodies, often in response to what they sense in their environment. For food, animals ingest plants and other organisms. The scientific study of animals is called zoology.

Animals have existed for millions of years, but it is not known when they first appeared on Earth. The earliest animals were soft-bodied, multicellular life-forms that did not preserve well as fossils. (A fossil is the remains or print of an organism from long ago that has been preserved in rock.) By the time animal parts became hardened in rock about 640 to 670 million years ago, numerous well-developed multicellular animals already existed. Therefore, the beginnings of the animal kingdom must have occurred earlier.

Most zoologists recognize the existence of 30 to 35 phyla (related groups) of animals, some of which are extinct (no longer exist) and are known only from their fossil record. Animals that live today come in many forms and sizes, the very smallest visible only under a microscope and the very largest, the blue whale, reaching 100 feet (30 meters) in length and weighing up to 300,000 pounds (136,000 kilograms). Animals are classified as vertebrates (having backbones) or invertebrates (without backbones).

Among the most primitive of the animals are the sponges, invertebrates that live in water. Sponges have no nerve cells or muscles. They are shaped like vases, with water flowing in through holes in their sides and leaving through an opening in their top. Sponges are remarkable for their ability to regenerate. Although many invertebrates are capable of growing new body parts, sponges are capable of growing into a new individual from even the tiniest fragment of the original body. It is believed that although sponges have lived successfully for about one billion years, yet it seems that they did not give rise to any other animal forms.

Some of the simplest animals with the oldest ancestral lines are invertebrates such as jellyfish, sea anemones and corals, which also live in water. They have radial body symmetry, which means that their bodies are arranged equally around a central point. This arrangement allows them to sense food and danger approaching from all directions. The simplest animals having bilateral symmetry (meaning that both the left and right sides of their bodies are mirror images of each other) include the invertebrate flatworms and roundworms. These animals live in water, on land, and as parasites in the body fluids of other organisms. Bilateral symmetry was an important evolutionary development because it allowed for forward movement of animals. It also is associated with the development of separate head and tail areas, as well as a distinction between the upper and under portions of an animal's body.

Another major step in the evolution of animals was the development of a body cavity called the coelom. The coelom is a cavity in the body between the gut and the body wall that houses the internal organs, such as the liver, stomach, and heart. Animal groups that have coeloms include Mollusca (snails, clams, octopus, and squid), Annelida (earthworms and leeches), Arthropoda (insects, spiders, and crabs), Echinodermata (sea urchins and starfish), and Chordata (fish, amphibians, reptiles, birds, and mammals). All of the vertebrates are included in the group Chordata.

About one million species of animals have been named. However, biologists estimate that a much larger number of animal species has yet to be discovered; the actual total could be as large as 30 to 50 million species.

References

- Jablonski D (1999). "The future of the fossil record". Science. 284 (5423): 2114–16. doi:10.1126/science.284.5423.2114. PMID 10381868
- Bayrakdar, Mehmet (1986). "Al-Jahiz and the rise of biological evolution"(PDF). Ankara Üniversitesi İlahiyat Fakültesi Dergisi. Ankara Üniversitesi İlahiyat Fakültesi. 27 (1): 307–15. doi:10.1501/Ilhfak_0000000674. Retrieved 21 December 2012
- Vassiliki Betty Smocovitis (1996). Unifying Biology: The Evolutionary Synthesis and Evolutionary Biology

Princeton University Press. ISBN 0-691-03343-9

- Black, J (Jun 2002). "Darwin in the world of emotions" (Free full text). Journal of the Royal Society of Medicine. 95 (6): 311–3. doi:10.1258/jrsm.95.6.311. ISSN 0141-0768. PMC 1279921 . PMID 12042386
- Wiley, R. H. (1981). "Social structure and individual ontogenies: problems of description, mechanism, and evolution" (PDF). Perspectives in ethology. 4: 105–133. Retrieved 21 December 2012
- Woese C, Kandler O, Wheelis M (1990). "Towards a natural system of organisms: proposal for the domains Archaea, Bacteria, and Eucarya". Proc Natl Acad Sci USA. 87 (12): 4576–9. Bibcode:1990PNAS...87.4576W. doi:10.1073/pnas.87.12.4576. PMC 54159 . PMID 2112744
- Heather Silyn-Roberts (2000). Writing for Science and Engineering: Papers, Presentation. Oxford: Butterworth-Heinemann. p. 198. ISBN 0-7506-4636-5
- Chambers, Geoffrey K.; Curtis, Caitlin; Millar, Craig D.; Huynen, Leon; Lambert, David M. (2014-01-01). "DNA fingerprinting in zoology: past, present, future". Investigative Genetics. 5: 3. doi:10.1186/2041-2223-5-3. ISSN 2041-2223. PMC 3909909 . PMID 24490906

Chapter 2

Vertebrate Zoology

Animals that have a backbone, such as reptiles, amphibians, fish, mammals and birds, are called vertebrates. They are studied under the domain of vertebrate zoology. Mammalogy, ornithology, herpetology and ichthyology are some of the significant subdisciplines of vertebrate zoology, which have been elaborated in this chapter.

Vertebrate zoology is the biological discipline that consists of the study of Vertebrate animals, i.e., animals with a backbone, such as fish, amphibians, reptiles, birds and mammals.

Vertebrate

Vertebrates are animals that are relatively complex, with highly developed anatomical and physiological features and complex behaviours. The same might be said for insects and some other animals; but vertebrates are different because they have a "notachord" and nerve chord.

As a rule, chordates are active animals. Their notochord is an exceptionally important characteristic of chordates. It is like a stiffened rod that does not compress. This allows the body muscles to act against the notochord and thus allows the animal to move.

Divisions of Vertebrate Zoology

The Division of Vertebrate Zoology focuses on animals with backbones -- fish, amphibians, reptiles, mammals, and birds. As a result of a historical legacy of large collecting expeditions to other continents during the last century (e.g., the AMNH Congo Expeditions, Archbold Expeditions to New Guinea, etc.) and ongoing field research expeditions by current staff and students, VZ houses one of the greatest collections of vertebrate specimens in the world. Over 3.5 million specimens representing over 35,000 species are archived in our collections, ranging from tiny fish and lizards to ostriches, elephants, and even whales. VZ collections also contain such rare and extinct species such as coelocanths, dodos, and Tasmanian wolves. Most of our specimens are traditional preparations (skins, skulls, skeletons, alcohol-preserved specimens) but we also maintain a growing collection of tissue samples for DNA studies (34,500). Staff and students in the Division of Vertebrate Zoology are engaged in research in diverse areas including systematics, phylogenetics, evolutionary morphology, population genetics, faunal inventory, biogeography, distribution modeling, conservation, host-parasite interactions, and extinctions.

The Division of Vertebrate Zoology is organized into four departments: Herpetology (amphibians and reptiles), Ichthyology (fishes), Mammalogy(mammals), and Ornithology (birds).

This subdivision of zoology has many further subdivisions, including:

- Ichthyology - the study of fishes.
- Mammalogy - the study of mammals.
- Ornithology - the study of birds.
- Herpetology - the study of reptiles.
- Batrachology - the study of amphibians.

Ichthyology

Ichthyology, scientific study of fishes, including, as is usual with a science that is concerned with a large group of organisms, a number of specialized subdisciplines: e.g., taxonomy, anatomy (or morphology), behavioral science (ethology), ecology, and physiology. Because of the great importance of fishes as human food, economic ichthyology is a significant segment of the field.

The ancient Greek naturalists, especially Aristotle, were highly knowledgeable about many aspects of fish biology, and until the end of the 19th century the main advances were in taxonomy, as new species were described and the relationships of those already known were clarified. In the mid-20th century a great increase in interest in oceanography, together with new techniques and equipment for underwater observations (especially the self-contained underwater breathing apparatus, or SCUBA), opened many new avenues for the study of fish behaviour and ecology under natural conditions. The development of improved methods of keeping fishes in tanks led to an increase in their use as laboratory animals in the fields of behaviour, ecology, functional anatomy, and even toxicology and parasitology.

A great variety of fishes are kept by aquarium hobbyists, a few of whom make valuable observations on fish behaviour. The majority of fish research, however, is conducted at institutional aquariums, in which large tanks can simulate natural conditions, and at universities and museums, which maintain large collections of preserved specimens. Much fishery research is conducted at the laboratories of government agencies, which are responsible for managing natural fish populations so as to maintain them as a renewable resource.

The study of fish, which is centuries old, reveals humanity's strong and lasting curiosity about nature, with fish providing both inner joy (beauty, recreation, wonder, and religious symbolism) and practical values (ecology, food, and commerce).

With about 27,000 known living species, fish are the most diverse group of vertebrates, with more than one-half of the total vertebrate species. While a majority of species have probably been discovered and described, approximately 250 new species are officially described by science each year.

Hagfish, while generally classified in Agnatha and as fish, actually lack vertebrae, and for this reason sometimes are not considered to be fish. Nonetheless, they remain a focus of ichthyology. Many types of aquatic animals named "fish," such as jellyfish, starfish, and cuttlefish, are not true fish. They, and marine mammals like whales, dolphins, and pinnipeds (seals and walruses) are not a focus of ichthyology.

The practice of ichthyology is associated with aquatic biology, limnology, oceanography, and aquaculture.

Early Developments

Fish compose approximately 8% of all figurative depictions on Mimbres pottery.

The study of fish likely receives its origins from the human desire to feed, clothe, and equip themselves with useful implements. Early ichthyologists were likely hunters and gatherers who investigated which fish were edible, where they could be found, and how to best capture them. These insights of early cultures were manifested in abstract and identifiable artistic expressions.

Around 3,500 B.C.E. the Chinese were trying to learn about fish in order to practice aquaculture. When the waters lowered after river floods, some fishes, namely carp, were held in artificial lakes. Their brood were later fed using nymphs and feces from silkworms used for silk production.

Moses, in the development of the kashrut (Jewish dietary laws), forbade the consumption of fish without scales or appendages. This required some study of fish that has continued to this day.

The Development of Modern Ichthyology

The writings of three sixteenth century scholars, Hippolyte Salviani, Pierre Belon, and Guillaume Rondelet, signify the conception of modern ichthyology. The investigations of these individuals were based upon actual research in comparison to ancient recitations. Despite their prominence, Rondelet's De Piscibus Marinum is regarded as the most influential, identifying 244 species of fish.

The incremental alterations in navigation and shipbuilding throughout the Renaissance marked the commencement of a new epoch in ichthyology. The Renaissance culminated with the era of exploration and colonization, and upon the cosmopolitan interest in navigation came the specialization in naturalism.

Georg Marcgrave of Saxony composed the Naturalis Brasilae in 1648. This document contained a description of one hundred species of fish indigenous to the Brazilian coastline. In 1686 John Ray and Francis Willughby collaboratively published Historia Piscium, a scientific manuscript containing 420 species of fish, 178 of these newly discovered. The fish contained within this informative literature were arranged in a provisional system of classification.

The classification used within the Historia Piscium was improved upon by Carolus Linnaeus (1707–1778), the "father of modern taxonomy." His two prime contributions were: (1) to establish conventions for the naming of living organisms using binomial nomenclature (the genus name followed by the species name), and (2) developing a hierarchical system for classification of organisms. Although the system now known as binomial nomenclature was developed by the Bauhin brothers (Gaspard Bauhin and Johann Bauhin) almost two hundred years earlier, Linnaeus was the first to use it consistently, and may be said to have popularized it within the scientific community. Linnaeus's taxonomic approach became the systematic approach to the study of organisms, including fish.

It was one of Linnaeus's colleagues, Peter Artedi, who earned the title "father of ichthyology" through his indispensable advancements. Artedi contributed to Linnaeus's refinement of the principles of taxonomy. Furthermore, he recognized five additional orders of fish: Malacopterygii, Acanthopterygii, Branchiostegi, Chondropterygii, and Plagiuri. Artedi developed standard methods for making counts and measurements of anatomical features that are modernly exploited. Another associate of Linnaeus, Albertus Seba, was a prosperous pharmacist from Amsterdam. Seba assembled a cabinet, or collection, of fish. He invited Artedi to utilize this assortment of fish; unfortunately, in 1735, Artedi fell into an Amsterdam canal and drowned at the age of 30. Linnaeus posthumously published Artedi's manuscripts as Ichthyologia, sive Opera Omnia de Piscibus (1738).

Linnaeus revised the orders introduced by Artedi, placing significance on pelvic fins. Fish lacking this appendage were placed within the order Apodes; fish containing abdominal, thoracic, or jugular pelvic fins were termed Abdominales, Thoracici, and Jugulares respectively. However, these alterations were not grounded within the evolutionary theory. Therefore, it would take over a century until Charles Darwin would provide the intellectual foundation from which it would be perceived that the degree of similarity in taxonomic features corresponded to phylogenetic relationship.

Modern Era

Close to the dawn of the nineteenth century, Marcus Elieser Bloch of Berlin and Georges Cuvier of Paris made an attempt to consolidate the knowledge of ichthyology. Cuvier summarized all of the available information in his monumental Histoire Naturelle des Poissons. This manuscript was published between 1828 and 1849 in a 22 volume series. This documentation contained 4,514 species of fish, 2,311 of these new to science. This piece of literature still remained one of the most ambitious treatises of the modern world.

The scientific exploration of the Americas advanced knowledge of the remarkable diversity of fish. Charles Alexandre Lesueur, a student of Cuvier, made a collection of fish dwelling within the Great Lakes and Saint Lawrence River regions.

Adventurous individuals such as John James Audubon and Constantine Samuel Rafinesque figure into the faunal documentation of North America. These persons often traveled with one another and composed Ichthyologia Ohiensis in 1820. In addition, Louis Agassiz of Switzerland established his reputation through the study of freshwater fish and organisms and the pioneering of paleoichthyology. Agassiz eventually immigrated to the United States and taught at Harvard University in 1846.

Albert Günther published his Catalogue of the Fishes of the British Museum between 1859 and 1870, describing over 6,800 species and mentioning another 1,700. Generally considered one of the most influential ichthyologists, David Starr Jordan wrote 650 articles and books on the subject as well as serving as president of Indiana University and Stanford University.

Today, ichthyology is a well-known scientific field investigating such areas related to fish as classification, anatomy, evolution, genetics, ecology, physiology, and conservation.

Fish

A fish is a vertebrate (an animal with a backbone) that lives in the water and breathes through gills. It has a streamlined body and is usually protected by a hard coat of scales. Fish species vary greatly in size and shape, but all are cold-blooded and most have fins. Fish live in watery habitats as diverse as stagnant ponds and subzero polar water.

As a vertebrate that lives its entire life underwater, a fish is ectothermic or cold-blooded. This means that it is not necessarily cold but that its own, internal temperature rises or lowers to meet that of its environment. If anything defines a fish as a fish, it is that it breathes through gills. These respiratory organs that lie behind and to the side of the mouth are able to absorb oxygen that is dissolved in water and to give off carbon dioxide as the water passes over the gills' filaments. Oxygen-rich blood is then pumped by a heart to the rest of its body.

Most fish have a streamlined body over which water flows easily as it moves through the water. Fish are able to swim forward by contracting the muscles on each side of their body in turn so that their tail whips from side to side and pushes them forward. Fins allow them to maneuver and have control and balance, while an inflatable swim bladder keeps them from sinking when they are not swimming. As a fish descends to deeper waters, the increase in pressure compresses and deflates the bladder, allowing the fish to swim deeper. As the fish rises again and the pressure decreases, the bladder begins to inflate with gas. The majority of fish have scales that are overlapping plates that protect its body. A fish does not shed these scales, because they grow as its body grows. Mucus usually covers these scales, as it helps the fish to glide more easily through the water.

Most fish reproduce sexually through the union of male sperm and female eggs, but it takes place outside the female's body by what is called spawning. Spawning is the release of eggs by the female into the water. The male then releases his sperm over the eggs and some of the eggs are fertilized.

Biologists have grouped fish into three classes: jawless fish, cartilage fish, and bony fish. A jawless fish is a primitive, wormlike fish without a hinged jaw. This means that it usually has a simple, sucker-like mouth instead. A lamprey eel is a good example of this ancient type of fish. The lamprey is a parasite and sucks the blood and juices from live fish. The only other jawless fish is the scavenger fish called the hagfish. It has a round mouth and attaches itself to the bodies of dead or dying fish, feeding on the contents of the victim's body.

Cartilage fish have an endoskeleton (internal skeleton) made entirely of strong, flexible cartilage instead of bone. These fish are almost always hunters such as sharks, skates, and rays who live in the ocean. They all have jaws, scales, and paired fins. Their skin is covered with tiny scales that feel like sandpaper. The shark is an especially ferocious predator that must swim all the time because it has no swimbladder. Sharks are powerful swimmers and have teeth that grow in rows that move forward to replace lost teeth.

Bony fish make up the third and largest category of fish. Also called "true fish," these have an endoskeleton made up mostly of bone. The familiar trout, salmon, cod, and sardine are all bony fish. Bony fish all have a gill cover or flap and like all fish, a keen sense of smell. Bony fish usually have highly maneuverable fins that allow them to make rapid and complex movements. Although bony fish are, as their name implies, all bone, their skeleton is in fact very light and thin because they use the natural buoyancy of water to support their bodies. Fish are an important food source for people.

Terminology: "Fish" vs. "Fishes"

Both "fish" and "fishes" are acceptable plurals, and both forms are common. For example, the later usage can be found throughout Nelson's *Fishes of the World, 3rd Edition* (1994), and within the *Guidelines for Use of Fishes in Field Research* (1988) by the American Society of Ichthyologists and Herpetologists, the American Fisheries Society, and the American Institute of Fisheries Research Biologists. Examples of the use of *fish* as the plural are quite common in popular literature.

Although both plural forms are acceptable, a common preference among biologists is to use the term "fishes" when speaking about two or more kinds of fish (species), and the term "fish" when referring to two or more individual fish organisms. For example, the publication Guidelines of the American Fisheries Society (2005) offers a preference for "fishes" as a synonym for "species of fish." Thus, in the statement "there are over 27,000 fishes in the world," the reference would be to over 27,000 fish species in the world. In the statement, "there are several million fish of the species *Gadus morhua*," the reference is that *G. morhua* comprises several million individuals. The usage of the two words is similar to that of the words "people" and "peoples." Of course, languages change over time and one may see a trend in the use of these terms as well.

Characteristics of Fish

Fish range in size from the 14m (45ft) whale shark to the 7mm (just over 1/4 of an inch) long stout infantfish and the 13mm Philippine goby. Fish can be found in almost all large bodies of water in salt, or brackish, or fresh water, at depths from just below the surface to several thousand meters. However, hyper-saline lakes like the Great Salt Lake of the United States do not support fish. Some species of fish have been specially bred to be kept and displayed in an aquarium, and can survive in the home environment.

Hagfish, while generally classified in Agnatha ("jawless") and as fish, actually lack vertebrae. For this reason, hagfish, which are also commonly known as "slime eels," are sometimes not considered to be fish. The other living member of Agnatha, the lamprey, has primitive vertebrae made of cartilage. Hagfish are a staple food in Korea. They are classified in the order Myxini and the family Myxinidae. Both hagfish and lamprey have slimy skin without scales or plates. They also have a notochord that remains throughout life; circular, jawless mouths; and unpaired fins. Hagfish are found in the oceans and lampreys are found in both freshwater and ocean environments. Most lampreys are parasitic.

Fish belonging to the class Chondrichthyes are distinguished by cartilage skeletons, as opposed to skeletons of bone. They have movable jaws and paired fins. Almost all of the Chondrichthyes—sharks, rays, skates, and chimaeras—are found in ocean environments.

Northern pike, a ray-finned fish

Most fish species (about 95 percent) are placed in the class Osteichthyes (which some taxonomies consider a superclass). They have bony internal skeletons and skins with scales. (As a general rule for the taxon. Not all bony fish have scales, and scales may be absent or present in two closely related species (Nelson 1994); Catfish is an example of an order of fish that lack scales.) The Osteichthyes taxon include coelacanths (lobe-finned fish), lungfish, and ray-finned fish. Coelacanths were thought to have been extinct for millions of years until fishermen caught one in 1938. Lungfish have lungs, as well as gills. Ray-finned fish are what many people refer to when they use the term fish, as these are our most familiar fish, including bass, eels, and many sports fish. Ray-finned fish have fins that are supported by bones (rays).

Many types of aquatic animals named "fish," such as jellyfish, starfish, and cuttlefish, are not true fish. A number of sea dwelling creatures, like dolphins and whales, are actually mammals.

While fish are poikilothermic in that they do not maintain constant internal temperatures and the temperature often mirrors the ambient temperature, certain species of fish maintain elevated body temperatures to varying degrees. These include teleosts (bony fishes) in the suborder Scombroidei and billfishes, tunas, and one species of "primitive" mackerel (*Gasterochisma melampus*). All sharks in the family Lamnidae—shortfin mako, long fin mako, white, porbeagle, and salmon shark—are known to have this capacity, and evidence suggests the trait exists in family Alopiidae (thresher sharks). The degree of being able to have elevated temperatures varies from the billfish, which warm only their eyes and brain, to bluefin tuna and porbeagle sharks, which can elevate body temperatures in excess of 20 °C above ambient water temperatures. In many cases, this phenomena has been traced to heat exchange, as warmer blood being returned to the gills in small veins runs close to colder, oxygenated blood in narrow arteries leaving the gills. This ability to have elevated temperatures allows fish to be active in colder waters and to have enhanced swimming ability because of the warmer muscles. In general, most fish can survive only at a relatively small range of body temperatures, but may adjust their depth in large bodies of water in order to find preferable ranges.

Classification

Fish are a very diverse assemblage, so much so that the term *fish* is itself more one of convenience than a taxonomic rank (Nelson 1994). It is used to designate a paraphyletic group, whereby the most recent common ancestor is included but not all descendants, with tetrapods (four-legged vertebrates, or descendants of four-legged vertebrates) being excluded. It is thus not surprising that different taxonomists may classify fish differently.

Latimeria chalumnae (coelacanth, a lobe-finned fish)

Vertebrates are generally classified into two groups, the Agnatha (jawless vertebrates) and the Gnathostomata (jawed vertebrates). The later group includes fish with hinged jaws, but also includes amphibians, reptiles, birds, and mammals (the tetrapods). In most taxonomies, Agnatha and Gnathostomata are each considered a *superclass*, although sometimes Agnatha is considered a *class*.

The Agnatha, in addition to including the modern day lampreys (Petromyzontiformes) and hagfish (Myxiniformes), also includes several extinct orders.

Within Gnathostomata, several classes of fish are recognized. Two of these classes have living representatives, the Chondrichthyes (cartilaginous fish) and Osteichthyes (bony fish). In some taxonomies, Osteichthyes is considered a superclass.

Within the Osteichthyes, two extant subclasses (or classes) are generally recognized, the Sarcopterygii (lobe-finned fish) and the Actinopterygii (ray-finned or spiny-finned fish). The coelacanths are generally placed within the Sacropterygii subclass. The Actinopterygii are generally divided into the Chondrostei and the Neopterygii, the latter of which includes the Teleostei (modern bony fishes), a classification into which most fish fit today.

A general grouping of fish, without reference to the names of ranks of *taxa* (superclass, class, subclass, etc.) is presented above, in the image box. Below is presented a more detailed taxonomic scheme with the rank names, as derived from that offered by Benton (2004), in his text *Vertebrate Paleontology*:

- SUBPHYLUM VERTEBRATA
  - Class 'Agnatha'
    - Subclass Myxinoidea (hagfish)
    - Subclass Petromyzontida (lampreys)
    - Subclass †Conodonta
    - Subclass †Pteraspidomorphi
    - Order †Thelodonti
    - Order †Anaspida
    - Subclass unnamed
      - Order †Galeaspida
      - Order †Pituriaspida
      - Order †Osteostraci
  - Infraphylum Gnathostomata (jawed vertebrates)
    - Class †Placodermi
    - Class Chondrichthyes (cartilaginous fish)
      - Subclass Elasmobrachii (sharks, rays, skates)
      - Subclass Subterbranchialia
        - Superorder Holocephali (chimaeras)
    - Class †Acanthodii
    - Class Osteichthyes (bony fish)
      - Subclass Actinopterygii (ray-finned fish)
        - Superdivision Chondrostei (sturgeons, paddlefish)
        - Superdivision Neopterygii (teleosts—modern bony fish)
    - Subclass Sarcopterygii (lobe-finned fish)
      - Order Dipnoi (lungfish)
      - Infraclass Crossopterygii
        - Order Actinistia (coelacanths)

† = extinct (the group lacks any living members)

Each year, biologists find 200 to 300 species of fish that had not been previously known (Grady 2006).

Fish and Humans

Throughout history, humans have utilized fish as a food source. Historically and today, most fish protein has come by means of catching wild fish. However, aquaculture, or fish farming, which has been practiced since about 3,500 B.C.E. in China, is becoming increasingly important in many nations. Overall, about one-sixth of the world's protein is estimated to be provided by fish. That proportion is considerably elevated in some developing nations and regions heavily dependent on the sea. In a similar manner, fish have been tied to trade. One of the world's longest lasting trade histories is the trade of dry cod from the Lofoten area in northern Norway to the southern parts of Europe. This trade in cod has been going on for more than 1000 years.

Fish are also caught for sport. Indeed, in many aquatic environments today, including most freshwaters, there are more fish caught for sport than via commercial fisheries.

Catching fish for the purpose of food or sport is known as fishing, while the organized effort by humans to catch fish is called a fishery. Fisheries are a huge global business and provide income for millions of people. The annual yield from all fisheries worldwide is about 100 million tons, with popular species including herring, cod, anchovy, tuna, flounder, and salmon. However, the term fishery is broadly applied, and includes more organisms than just fish, such as mollusks and crustaceans, which are often called "fish" when used as food.

Fish have been recognized as a source of beauty for almost as long as used for food, appearing in cave art, being raised as ornamental fish in ponds, and displayed in aquariums in homes, offices, or public settings. As of 2006, there were an estimated 60 million aquarium enthusiasts worldwide.

Basking shark, the second largest living fish (after the whale shark) is a filter feeder that eats zooplankton.

Because of the popularity of fish for food, sport, and hobby, overfishing is a threat to many species of fish. It was reported that all large oceanic fish species worldwide had been so systematically overcaught that fewer than 10 percent of 1950 levels remained. Particularly imperiled were sharks, Atlantic cod, Bluefin tuna, and Pacific sardines.

Some fish pose dangers to humans. Although the sharks may be among the most feared, there are actually few shark species that are known to attack humans. The largest sharks, the whale shark and basking shark, are actually plankton feeders. The International Shark Attack File reports there are only about 10-15 deaths each year worldwide. This compares to about 1,000 deaths annually from crocodiles and 60,000 from snakebites.

On the other hand, Smith and Wheeler (2006) suggest that, in contrast to prior estimates of 200 venomous fishes, 1,200 species of fish should be presumed venomous. Most of these venomous fishes come from off the coast of eastern and southern Africa, Australia, Indonesia, Phillipines, Polynesia, and southern Japan (Grady 2006). About 50,000 people a year suffer from fish stings or envenomations (Grady 2006). Perhaps the most dangerous venomous fish is the stonefish, which can release a venomous toxin from spikes on its back when it is provoked or frightened. This toxin can be fatal to humans if not treated promptly. The pufferfish, often better known by the Japanese name Fugu, poses risks to humans because this species contains a highly toxic poison in the internal organs. Despite this, it is considered a delicacy in Japan. The pufferfish needs to be very specially prepared to be safe for eating. Every year a number of people die from consuming this fish.

Barracudas, sea bass, moray eels, and stingrays are among other fish that pose risks to humans in the aquatic environment.

Fish have a prominent role in in human religious culture, particularly Christianity. The fish is used often as a symbol by Christians to represent Jesus or Christianity. The Greek word for fish (ichthus, spelled Iota Chi Theta Upsilon Sigma) is taken as an acrostic, spelling out Jesus and his titles: "Jesus Christ, Son of God, Savior" (*Iesous CHristos THeou Uiou Soter*), or similar combinations. The gospels of the New Testament contain many references to fishing, both literal, as when Jesus directed Peter where to cast his nets, and metaphorical, as when Jesus is recorded as stating, "Follow me, and I will make you fishers of men" (Matthew 4:19).

Types of Fish

Fish, the oldest vertebrate group, includes a huge range of types, from the Middle Ordovician, about 490 million years ago, to the present day. These are the main groups:

- Agnatha: the jawless fish. Cambrian to present day.
    - Pteraspids: the head-shields
    - Anaspids: gills opened as holes. Silurian to end-Devonian.
        - Cephalaspids: early jawless fish
        - Lampreys: living ectoparasites
    - Osteostraci: bony-armoured jawless fish.
- Gnathostomata: the jawed fish. Includes all types commonly called fish, except the lamprey.
    - Placoderms: heavily armoured fish

    - Chondrichthyes: cartilaginous fish: sharks, rays and skates.
    - Acanthodii: extinct spiny sharks
- Osteichthyes: bony fish.
    - Actinopterygii: the ray-finned fish.
        - Chondrostei: sturgeons and some other early types.
        - Neopterygii: first seen in the later Permian, lighter and faster-moving than previous groups.
        - Holostei: the gars and bowfins
        - Teleostei: the most successful group, Triassic to present day.
    - Sarcopterygii: the lobe-finned fish
        - Dipnoi: the lungfish; eight genera survive.
        - Coelacanths: two species survive. They were probably a sister-group to the tetrapods.

Certain animals that have the word *fish* in their name are not really fish: Crayfish are crustaceans, and jellyfish are Cnidarians. Some animals look like fish, but are not. Whales and dolphins are mammals, for example.

Anatomy

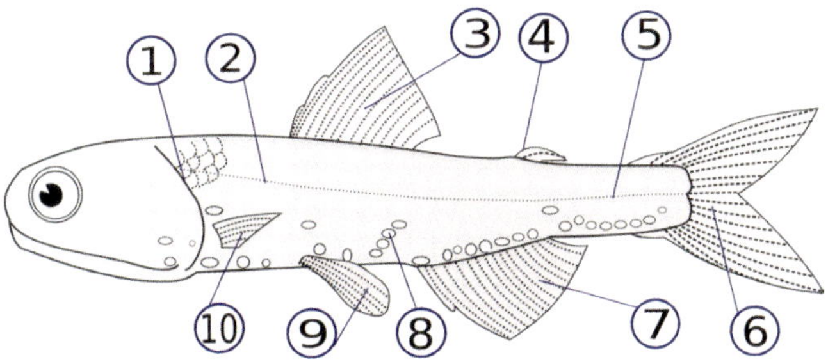

The anatomy of *Lampanyctodes hectoris*
1. operculum (gill cover) 2 & 5. lateral line 3. dorsal fin 4 fat fin 6. caudal fin 7. anal fin 8. photophores 9. pelvic fins paired 10. pectoral fins (paired)

Bony and Cartilagenous Fish

Most kinds of fish have bones. Some kinds of fish, such as sharks and rays, do not have real bones (their skeletons are made of cartilage) they are known as cartilaginous fish.

Fish Scales

All fish are covered with overlapping scales, and each major group of fish has its own special type of scale. Teleosts ('modern' fish) have what are called *leptoid* scales. These grow in concentric circles and overlap in a head to tail direction like roof tiles. Sharks and other chondrichthyes have *placoid* scales made of denticles, like small versions of their teeth. These also overlap in a head to tail direction, producing a tough outer layer. Shark skin is available for purchase as shagreen, a leather which as original is smooth in one direction, and rough in the other direction. It may be polished for use, but is always rough in texture and resistant to slipping.

The scales are usually covered with a layer of slime which improves passage through the water, and makes the fish more slippery to a predator.

Swimming

Fish swim by exerting force against the surrounding water. There are exceptions, but this is usually done by the fish contracting muscles on either side of its body in order to generate waves of flexion that travel the length of the body from nose to tail, generally getting larger as they go along. Most fishes generate thrust using lateral movements of their body & tail fin (caudal fin). However, there are also species which move mainly using their median and paired fins. The latter group profits from the gained manoeuvrability that is needed when living in coral reefs for example. But they can not swim as fast as fish using their bodies & caudal fins.

Muscle

Fish can swim slowly for many hours using red muscle fibres. They also make short, fast bursts using white muscle. The two types of muscle have a fundamentally different physiology. The red fibres are usually alongside a much greater number of white fibres.

The white fibres get their energy by converting the carbohydrate glycogen to lactate (lactic acid). This is anaerobic metabolism, that is, it does not need oxygen. They are used for fast, short bursts. Once the lactic acid builds up in the muscles, they stop working, and it takes time for the lactate to be removed, and the glycogen replaced. Using their white fibres, fish can reach speeds of 10 lengths per second for short bursts.

Swimming for long periods needs oxygen for the red fibres. The oxygen supply has to be constant because these fibres only operate aerobically. They are red because they have a rich blood supply, and they contain myoglobin. Myoglobin transports the oxygen to the oxidising systems. Red muscle gets its energy by oxidising fat, which weight for weight has twice as much energy as carbohydrate or protein. Using their red fibres, fish can keep up a speed of 3–5 lengths per second for long periods.

Swimming in Groups

Many fish swim in groups. Schools of fish can swim together for long distances, and may be chased by predators which also swim in schools. Casual groups are called 'shoals'.

Body Shape

The shape of the body of a fish is important to its swimming. This is because streamlined body shapes makes the water drag less. Here are some common fish shapes:

Streamlining

The rocket-shape of this shark makes it an efficient swimmer. It is fast over short distances.

The picture above shows a shark. This shark's shape is called *fusiform*, and it is an ovoid shape where both ends of the fish are pointy. This is the best shape for going through water quickly. Fishes with fusiform shapes can chase prey and escape predators quickly. Many live in the open ocean and swim constantly, like marlins, swordfish, and tuna. Ichthyosaurs, porpoises, dolphins, killer whales all have similar shapes. This is an example of convergent evolution.

Eel-like

The long, ribbon-like shape of an eel's body shows another shape. This enables them to hide in cracks, springing out quickly to capture prey, then returning quickly to their hiding spot.

Flatfish

An adult flounder has both eyes on the upper side of its body.

Flatfish live on the bottom of the ocean or lake. Most use camouflage: they change colors to match the ocean floor.

Compressed

Fish with compressed shapes have flat, vertical bodies, with one eye on each side. They swim upright and can be very thin. They usually live in reefs where their flat bodies can slip in and out among the corals, sponges, and rocks, keeping hidden from predators. Angelfish, surgeonfish, and butterflyfish are all compressed fish.

Fish as Food

Some people eat fishes. Many kinds of fishes are eaten. The fishes that people eat most include carp, cod, herring, perch, sardines, sturgeon, tilapia, trout, tuna, and many others. A person who buys and sells fishes for eating is called a fishmonger.

The word *to fish* is also used for the activity of catching fishes. People catch fishes with small nets from the side of the water or from small boats, or with big nets from big boats. People can also catch fishes with fishing poles and fishhooks with bait. This is often called angling. Anglers also different types of fishing lures.

Because people are catching too many fishes for food or other uses, there are less and less fishes in the sea. This is a problem known as Overfishing.

Fish as Pets

Selective breeding of carp made them into the domesticated koi in Japan, and goldfish in China. This breeding began over 2,000 years ago. The Chinese brought their goldfish indoors during the Song Dynasty. They kept them in large ceramic vessels. That we now do in glass fish tanks.

Koi (and goldfish) have been kept in decorative ponds for centuries in China and Japan

Mammalogy

Mammalogy is the branch of zoology that deals with mammals. Major subject areas in mammalogy include anatomy (structure), physiology (function), behavior, ecology, evolution, and classification. Humans are mammals and belong to the class Mammalia which is one of the most diverse groupings of animals.

Mammals are among the smartest, fastest, and largest animals on Earth. The cheetah is the fastest mammal; the blue whale is the largest; and human beings are the most intelligent. There are 18 orders of mammals, containing about 4,000 living species. Aside from a couple of exceptions, all mammals have certain things in common. They are warm-blooded (they maintain a constant internal temperature despite their environment), have hair on their bodies, give birth to live young, and feed their newborns with milk from their mammary glands. It is because of these milk glands that mammals got their name. Beginning with the Greek philosopher Aristotle (384–322 b.c.), the Greeks were the first to systematically study, categorize, and write about mammals. In fact, it was Aristotle who recognized that both whales and dolphins were really mammals and had more in common with land-based animals than they did with fish.

Mammalogy allows us to understand how and where a mammal lives, what are its habits and behavior, and how it reproduces. Aside from the common mammal features already mentioned, there are other characteristics of mammals that deserve mention. All mammals have a basic structure. They all have skulls that house a brain, and they all have seven vertebrae (bony segments) in their necks, whether their neck is as long as a giraffe's or as short as a dormouse. Most have four limbs that end in five digits (finger-like projections). Their teeth are adapted to their feeding habits, and most carry their unborn inside their bodies until birth. Mammals have highly developed senses, although not all are so sharp in every mammal. Some rely more on keen vision, while others depend most on their sense of smell. Most mammals are herbivorous (feeding on plants), while fewer are carnivorous (feeding on meat or other animals). Mammals are found in almost all habitats, and many actually build some sort of shelter or dwelling for themselves. For example, beavers build underwater lodges, gorillas make beds of palms, and prairie dogs have underground tunnels. Some mammals hibernate, or enter a sleeplike state, during winter, while others migrate or travel some distance to avoid winter weather.

While many mammals live alone as adults, many also live in groups of different sizes. For example, humans live in families typically consisting of a male, female, and their offspring while beavers live in family groups. Monkeys live in larger groupings called bands, while sheep live in larger groups called herds. The largest grouping of mammals are called colonies, and this is how bats live. As social animals, mammals need to communicate, and they do this in many different ways. Some communicate by a scent and give off a certain smell when they are in heat (and are ready to mate). Many use visual signals. For instance, while a gorilla may make certain facial gestures, a wolf will assume a certain body stance. Most mammals, however, use sounds to communicate, from a coyote's howl to a beaver's tail-slapping.

At the beginning of the twenty-first century, it is safe to say that nearly all of the world's mammal species are known to science. Yet the same cannot be said about the biology of every species. Twenty-first century techniques and technology will give mammalogists (people who study mammals) the ability to study free-living (animals in the wild) animals by the use of data obtained from tiny radio transmitters placed on the animals. These transmitters will also allow scientists to learn more about the genetics of mammals.

The study of mammals can be as diverse as the organisms themselves. A mammalogist might study a wide variety of topics on a particular species or group of mammals or might take a comparative approach and investigate one aspect with regard to a wide variety of mammals. The major subdivisions of the science of mammalogy include the following:

Natural History

The study of a species' way of life, including descriptions of the habitats in which it lives, the food it eats, the predators that eat it, the stages of its maturation and reproduction and its social structure.

Taxonomy and Systematics

The study of the classification of mammals into distinct orders, families, genera, species and subspecies, thereby defining the geographic distribution of each taxon (a taxonomic category or group, such as a phylum, order, family, genus or species). In addition, the evolutionary relationships among extinct (paleontology) and/or living taxa are analyzed.

Anatomy and Physiology

The study of mammalian body structures and tissues and how each functions.

Ethology

The study of the behavior of an animal and how those behaviors influence its survival and reproduction.

Ecology

The study of interactions between mammals and their environments (nonliving and living). This discipline includes the study of special adaptations to environmental factors (physiological ecology) and the study of interactions within and among species (population and community ecology).

Management and Control

The interactions between humans and other mammals in which humans manipulate either the environment or populations of mammals to favor the use and/or survival of certain species and to regulate or even reduce the populations of species whose activities conflict with human interests.

Of course, individual studies in mammalogy are not necessarily restricted to one of the above categories; many studies combine aspects from two or more areas. In game management, for example, the ecology and natural history of the species must be understood, and in taxonomy anatomical features are used to discern systematic relationships. In addition, mammalogists often collaborate with scientists in other fields, such as ornithologists in studies of predator-prey relationships between birds-of-prey and rodents and botanists in studies of plant-herbivore interaction. Obviously, becoming a mammalogist is not the only way to study mammals. In fact, many scientists who study mammals would not consider themselves mammalogists per se, but rather specialists in such fields as physiology or ethology who happen to use mammals as the subjects in their research. A number of employment opportunities exist for trained mammalogists and individuals interested in working with mammals. Each job differs in its duties/responsibilities, methodologies, perspectives, compensation and level of training required.

Mammals

Mammals are a well-known class of vertebrates, including many familiar domesticated species and pets, as well as our own species Homo sapiens. All mammals are warm-blooded, and all female mammals possess mammary glands (mammae), which are used to suckle the young with milk. Mammals are further distinguished by the possession of hair or fur, although this is limited to early developmental stages in the Cetacea (whales and dolphins). The vast majority of mammals give birth to live young, the exception being the egg-laying Monotremata (a small group of mammals including the duck-billed platypus and the echidnas or spiny anteaters). Monotremes are found exclusively in Australia and New Guinea. Mammals are found in a wide variety of habitats, including terrestrial, freshwater, and marine systems. They occur from the deserts to the dense forests, from the deep seas to the highest mountains, and from the tropics to the polar ice caps. Only one group of mammals, the Chiroptera (bats) is adapted to flight; other taxa such as flying squirrels or flying possums can glide effectively but are not capable of true flight.

Main Characteristics

Reproductive Cycle

Most marsupial and eutherian mammals have a reproductive cycle known as the oestrous cycle (U.S: estrous cycle). Females are sexually active only during the oestrous stage, when they are 'on heat' for a few days each month. If an ovum is not fertilized, the endometrium (uterus lining) is resorbed. Oestrus cycles may occur once or twice a year, or many times a year. Each group of mammals has its own frequency.

Humans and primates, are quite different. They have a menstrual cycle. In this case, females are sexually receptive at any time, but only fertile when an ovum is released from an ovary. In this case, the endometrium (if not needed for an fertilised egg) is discarded. The endometrium is shed, and takes with it a certain amount of blood. In this system, eggs are released from the ovaries mostly in the middle of the cycle, away from the menstrual period. This ovulation is 'concealed', meaning, it is not obvious when it occurs. This process, so it is thought, tends to keep the male and female together, which is unusual in mammals with the oestrous cycle.

Skeleton

One diagnostic feature is the lower jaw which, unlike earlier forms, is composed of a single bone, the dentary. This is one feature which can be seen in fossils, or at least those which are complete enough to have the lower jaw. Mammals have three little bones in their inner ear, the ear ossicles. The ear ossicles are bones which were, long ago, part of the lower jaw in early proto-mammals.

There are quite a number of other features, particularly in the skull and limbs, so that it is usually possible identify and describe a mammal from its skeleton alone.

Neocortex and Behavior

Another diagnostic feature is the neocortex of the brain, which no other vertebrate has. This is involved in the kind of flexible behavior and learning typical of mammals. Reptiles and birds have much of their behavior controlled by "inherited behavior chains", which roughly translates as "instincts". Almost all animals can do some learning, but mammals do far more than other vertebrates. Their behavior is much more flexible than lizards, for example, and that is made possible by their neocortex.

Other things in the life of mammals seem to be connected with this flexibility and learning. Play is a kind of early learning period in which, according to one theory, mammals develop skills which they will need in life. All mammalian young play, and this is very obvious in the more intelligent mammals (primates, cats).

The emotions of mammals are very noticeable, and rather similar to ours. It is possible, and quite common, for humans to have a friendly relationship with another mammal. It is quite impossible for a human to have any kind of relationship with a snake or a gecko (for example). This is because the reptile simply does not have the same basic emotions as a human.

Other Items

There are about 50 characters which are typical of mammals, and some of the most important are discussed above. A few more examples will make it clear that mammals are very different from reptiles and birds:

- Sweat glands
- Tooth replacement: two sets, and no continuous replacement. Enamel on the tooth surface. Reptile teeth all alike; mammal teeth follow a pattern (incisors, canines, premolars and molars)
- Occipital condyles. Two knobs at the base of the skull fit into the topmost neck vertebra; most tetrapods, in contrast, have only one such knob
- Outlet for food waste separate from urinogenital outlet. Reptiles and birds have a common cloaca at the rear
- Mammals excrete urea; reptiles and birds excrete uric acid
- Colour vision is defective or absent in most mammals (primates are an exception)
- In reptiles and birds the blood vessel which carries oxygenated blood from the left ventricle of the heart lies on the right side of the body; but in mammals it lies on the left side
- There are many features of the skeleton that mammals share
  - Their neck almost always has seven vertebrae, no matter how long it is.
  - Their lower jaw is made of just a single bone on each side, the dentary.
  - Their inner ear has three tiny bones, the ossicles: malleus, incus and stapes.

In the language of cladistics, these 50 unique characters are apomorphies which prove that mammals are a clade descended from a common ancestor.

Main Groups

Almost all mammal species give birth to live young. There are only two mammals that lay eggs, called monotremes, the duck-billed platypus *Ornithorhynchus*, and the spiny anteater *Echidna*, with four species. All are confined to Australia and New Guinea, and are the sole survivors of an earlier group of mammals. However, like other mammals, they feed milk to their young, and protect and look after them.

Other mammals are divided into the marsupials and the Eutheria, the placental mammals. Marsupials are mammals with pouches to carry young in, like the kangaroo.

Modes of Life

In terms of number of species mammals, with 5488 species, are not the most successful vertebrates. Birds, with about 10,000 species have nearly twice as many, and reptiles have just as many as birds. Fish have even more species. There are 27,000 species of fish, of which nearly 26,000 are bony fish.

Despite this, many zoologists do regard mammals as a successful group of animals. One reason is that they are successful in all habitats on Earth. In the air, in the water, in forests, in the colder regions of the world, and above all on grasslands, where they are outstandingly successful.

In the air, the bats (Chiroptera) are the mammalian order with the most species. They 'own' the nighttime, since birds are largely diurnal (daytime) animals. Bats are hugely successful, mostly as nighttime predators of insects.

Seagoing mammals, the Cetacea and the pinnipeds, are very successful and significant predators. This includes the whales, seals, walrus, dolphins and others.

The terrestrial mammals are fewer in number of species than lizards, but they are huge in individual numbers, and far more important in the life of the terrestrial biomes. Their ability to move from place to place and adapt has made them a most effective group. Many mammals live in cold places. These mammals have thick hair or blubber to keep them warm. Others may live in rainforests. On land the rodents (rats, mice) are hugely successful, more common in numbers than any other mammals. Large mammals on land have been hunted to extinction in some parts of the world, but the ones which remain are now better protected.

Last, but certainly not least, are the primates. Their natural habitat, with few exceptions, are the forests. Most live in the trees, with hands that grasp, good colour vision, and intelligence. In the Pliocene period some moved out onto the savannahs as grassland replaced forests. Mankind is the result of this shift into the savannahs.

Taxonomy

The evolutionary relationships among land vertebrates is as follows:

- Tetrapoda (land vertebrates)
    - Amphibia (lay eggs in water)
    - Amniota (lay cleidoic eggs)
        - Sauropsida (all true 'reptiles')
        - Synapsida (all 'mammal-like amniotes')
        - Pelycosauria

    - Therapsida (in particular, the Cynodontia)
    - Mammalia (bear live young; milk; raise their young etc.)

This sort of classification is not traditional, but it does reflect our knowledge of palaeontology and evolution.

Standardized Textbook Classification

A somewhat standardized classification system has been adopted by most current mammalogy classroom textbooks. It is based on living animals. The following taxonomy of extant and recently extinct mammals is from Vaughan et al. 2000.

Class Mammalia

- Subclass Prototheria: monotremes: platypuses and echidnas
- Subclass Theria: live-bearing mammals
    - Infraclass Metatheria: marsupials
    - Infraclass Eutheria: placentals

List of Orders

Mammals can be divided in a number of orders:

- Monotremes (monotremata)
- Marsupials (marsupialia)
- Eutheria
    - Superorder Xenarthra
        - Order Pilosa
        - Order Cingulata
    - Superorder Afrotheria
        - Order Proboscidea
    - Superorder Laurasiatheria
        - Insectivora
        - Order Eulipotyphla
        - Order Chiroptera
        - Order Carnivora
        - Order Perissodactyla
        - Order Artiodactyla

- Order Cetacea
  - Superorder Euarchontoglires
    - Order Rodentia
    - Order Lagomorpha
    - Order Primates

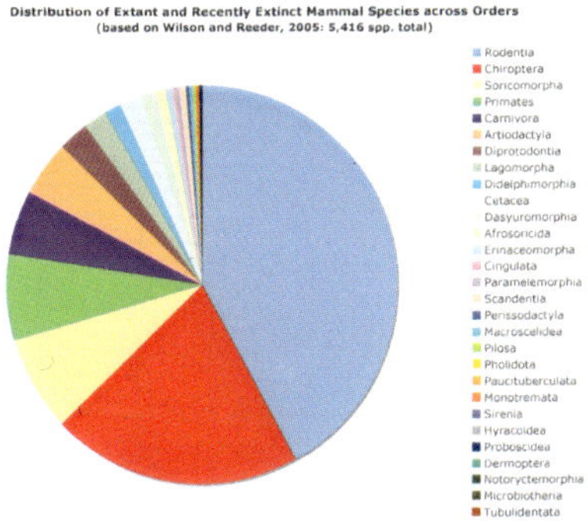

Over ¾ of mammal species are in the orders Rodentia (blue), Chiroptera (red), Soricomorpha (yellow) and Primates (green).

Debate on the Meaning of "Mammalia"

Because two quite different dates are given in the taxobox, an explanation is needed. Rowe defines mammals as "the taxon originating with the most recent common ancestor of extant [living] Monotremata and Theria". That puts the emphasis heavily on living mammals, where, as Rowe points out, the database of characters is extensive.

Kemp explains the problem with that approach: "If the definition of a mammal is based rigorously upon possession of all the characters of living mammals, then some fossil forms that are extremely mammalian in anatomy... are excluded".

"An altogether different perspective on defining Mammalia is based on traditional palaeobiological practice. An arbitrary decision is made about which characters to select as defining characters... Characters deemed appropriate are those reflecting the... fundamental mammalian biology. The essence of mammalian life is to be found in their endothermic temperature physiology, greatly enlarged brain, dentition capable of chewing food, highly agile, energetic locomotion, and so on. The organisms that achieved this grade of overall organisation are deemed to be Mammalia... Around the end of the Triassic period, about 205 mya, a number of fossils are found of very small animals that have [most] of the skeletal characters of modern mammals".

This difference in outlook explains the difference in the two dates given in the taxobox.

Habitat, Physiological Characteristics, and Behavior

Different species of mammals have evolved to live in nearly all terrestrial and aquatic habitats on Earth. Mammals inhabit every terrestrial biome, from deserts to tropical rainforests to polar icecaps. Many species are arboreal, spending most or all of their time in the forest canopy. One group (bats) has even evolved powered flight, only the third time that this ability has evolved in vertebrates (the other two groups being birds and extinct Pterosaurs).

Many mammals are partially aquatic, living near lakes, streams, or the coastlines of oceans (e.g., seals, sea lions, walruses, otters, muskrats, and many others). Whales and dolphins (Cetacea) are fully aquatic, and can be found in all oceans of the world, and some rivers. Whales can be found in polar, temperate, and tropical waters, both near shore and in the open ocean, and from the water's surface to depths of over 1 kilometer.

All mammals have hair at some point during their development, and most have hair their entire lives. Adults of some species lose most or all of their hair but, even in mammals like whales and dolphins, hair is present at least during some phase of ontogeny. Mammalian hair, made of a protein called keratin, serves at least four functions. First, it slows the exchange of heat with the environment (insulation). Second, specialized hairs (whiskers or "vibrissae") have a sensory function, letting an animal know when it is in contact with an object in its environment. Vibrissae are often richly innervated and well supplied with muscles that control their position. Third, hair affects appearance through its color and pattern, helping to camouflage predators or prey as well as signal to predators a defensive mechanism (for example, the conspicuous color pattern of a skunk is a warning to predators). Hair also communicates social information (for example, threats, such as the erect hair on the back of a wolf; sex, such as the different colors of male and female capuchin monkeys; or the presence of danger, such as the white underside of the tail of a white-tailed deer). Fourth, hair provides some protection, simply by providing an additional protective layer (against abrasion or sunburn, for example) or by taking on the form of dangerous spines that deter predators (porcupines, spiny rats, others).

Mammals are typically characterized by their highly differentiated teeth. Teeth are replaced just once during a mammal's life (a condition called diphyodonty). Other characteristics found in most mammals include: a lower jaw made up of a single bone, the dentary; four-chambered hearts; a secondary palate separating air and food passages in the mouth; a muscular diaphragm separating thoracic and abdominal cavities; a highly developed brain; endothermy and homeothermy; separate sexes, with the sex of an embryo being determined by the presence of a Y or 2 X chromosomes; and internal fertilization.

Development

There are three major groups of mammals, each united by a major feature of embryonic development. Monotremes (Prototheria) lay eggs, which is the most primitive reproductive condition in mammals. Marsupials (Metatheria) give birth to highly altricial young after a very short gestation period (8 to 43 days). The young are born at a relatively early stage of morphological development. They attach to the mother's nipple and spend a proportionally greater amount of time nursing as they develop. Gestation lasts much longer in placental mammals (Eutheria). During gestation, eutherian young interact with their mother through a placenta, a complex organ that connects the embryo with the uterus. Once born, all mammals are dependent upon their mothers for milk. Aside from these few generalities, mammals exhibit a diversity of developmental and life history patterns that vary among species and larger taxonomic groups.

Reproduction

Most mammalian species are either polygynous (one male mates with multiple females) or promiscuous (both males and females have multiple mates in a given reproductive season). Because females spend a long period of time in gestation and lactation, it is often the case that males can produce many more offspring in a mating season than can females. Thus, the most common mammalian mating system is polygyny, with relatively few males fertilizing multiple females and many males fertilizing none. This causes intense competition between males in many species, and the potential for females to be discriminating in choice of sire for offspring, leading to complex behaviors and morphologies associated with reproduction. Many mammal groups are marked by sexual dimorphism as a result of selection for males that can better compete for access to females.

About 3 percent of mammalian species are monogamous, males mating with a single female each season. In these cases, males provide at least some care to their offspring. Often, mating systems may vary within species depending upon local environmental conditions. For example, when resources are low, males may mate with a single female and provide care for the young. When resources are abundant, the mother may be able to care for young on her own and males will attempt to sire offspring with multiple females.

Other mating systems, such as polyandry, can be found among mammals. Some species (e.g. common marmosets and African lions) display cooperative breeding, in which groups of females, and sometimes males, share the care of young from one or more females. Naked mole rats have a unique mating system among mammals. Like social insects (Hymenoptera and Isoptera), naked mole rats are eusocial, with a queen female mating with several males and bearing all of the young in the colony. Other colony members assist in the care of her offspring and do not reproduce themselves.

Mammals range from many altricial young in each bout of reproduction (rodents and insectivores) to those species that give birth to one or a few precocial young. The former tend toward high early mortality and short lifespans, while the latter invest energy in a few offspring that develop into efficient competitors, living longer in generally stable environments (Cetaceans, primates and artiodactyls). Most mammals make use of a den or nest for the protection of their young. Some mammals, however, are born well developed and able to locomote on their own soon after birth. Most notable in this regard are artiodactyls such as wildebeest or giraffes. Cetacean young must also swim on their own shortly after birth.

Generally, smaller mammals live short lives and larger mammals live longer lives. Bats (Chiroptera) are an exception to this pattern, being relatively small mammals that can live for one or more decades in natural conditions, considerably longer than natural lifespans of significantly larger mammals. Mammalian lifespans range from one year or less to 70 or more years in the wild. Bowhead whales may live more than 200 years.

Behavior

There are mammal species that exhibit nearly every type of lifestyle, including fossorial, aquatic, terrestrial, and arboreal lifestyles. Locomotion styles are also diverse: mammals may swim, run, bound, fly, glide, burrow, or climb as a means of moving throughout their environment.

Social behavior and activity patterns vary considerably as well. Some mammals live in groups of tens, hundreds, thousands or more individuals. Other mammals are generally solitary except when mating or raising young. Mammals may be nocturnal, diurnal, or crepuscular.

Olfaction, hearing, tactile perception, and vision are all important sensory modalities in mammals. Olfaction plays a key role in many aspects of mammalian ecology, including foraging, mating and social communication. Mammalian hearing is well developed as well. In some species, it is the primary form of perception. Echolocation, the ability to perceive objects in the external environment by listening to echoes from sounds generated by an animal, has evolved in several groups, including microchiropteran bats (Chiroptera) and many toothed whales and dolphins (Odontoceti). Vision is well developed in a large number of mammals, although less important in many species that live underground or use echolocation.

Mammals can be carnivores (e.g., most species within Carnivora), herbivores (e.g., Perissodactyla, Artiodactyla), or omnivores (e.g., many primates). Mammals eat both invertebrates and vertebrates (including other mammals), plants (including fruit, nectar, foliage, wood, roots, and seeds) and fungi. Being endotherms, mammals require much more food than ectotherms of similar proportions. Thus, relatively few mammals can have a large impact on the populations of their food items.

Economic Impact on Humans

Mammals are a vital economic resource for humans. Many mammals have been domesticated to provide products such as meat and milk (e.g., cows and goats) or fiber (sheep and alpacas). Many mammals are kept as service animals or pets (e.g., dogs, cats, ferrets). Mammals are important for the ecotourism industry as well: people travel to zoos and throughout the world to see animals like elephants, lions, or whales. Bats often help control populations of crop pests. Norway rats and domestic mice are vitally important in medical and other scientific research, serving as models in human medicine and research.

On the other hand, some species have a detrimental impact on humans and the ecosystem. Many mammals that eat fruit, seeds, and other types of vegetation are crop pests. Carnivores are often a threat to livestock and even to human lives. Mammals common in urban or suburban areas have caused automobile accidents when straying into roads and have become household pests. A few species coexist exceptionally well with people, including some feral domesticated mammals (e.g., rats, house mice, pigs, cats, and dogs). As a result of intentional or unintentional introductions near human habitation, these animals have had considerable negative impacts on the local biota of many regions of the world, especially the endemic biota of oceanic islands.

Many mammals can transmit diseases to humans or livestock. The bubonic plague is perhaps the most well-known example, spread via fleas carried by rodents. Rabies, which can be transmitted among mammalian species, is also a significant threat to livestock and can kill humans as well.

Ornithology

Ornithology is the study of birds. It is a true science of course but it is also a way of life for many amateur bird watchers. Most ornithologists started out as bird watchers. A few lucky ones get to be paid for their passion. The rest of us can still enjoy this enjoyable pastime, studying and observing the birds of our world.

Nineteenth century ornithology was primarily about taxonomy and collecting specimens. This was Darwin's job on the Beagle. Although he also made observations about behaviour, his foremost job was collecting museum specimens, which were important for identification and descriptions of new species. They would also be pivotal in his development of the theory of Natural Selection as the mechanism for Evolution.

Luckily modern ornithology is not just about shooting birds and stuffing them. In fact, now that most of the world's birds have been identified and classified, the work of modern ornithologists is about gathering data vital to the conservation of birds in the face of habitat destruction and human interference in natural ecosystems. And of course it is also about increasing scientific knowledge about birds: where do birds live, what do they eat, how do they reproduce and care for their young? And all of these questions can be answered not only by professional ornithologists but also by 'Twitchers', bird watchers with a keen eye and a love for their subjects.

Ornithology is a specialised field but there is still room for bird watchers. In many countries yearly counts are made of bird species and this work is primarily carried out by 'twitchers'. Although they aren't paid for this, they can be very competitive, especially when seeing how many species can be identified in an area in a single 24 hour period. Twitchers also play an important role in conservation, often at the forefront of efforts to protect prime habitats for endangered species.

Another aspect of ornithology in the general sense is keeping birds as pets and this is an area that many bird lovers indulge in and which also creates quite a few jobs, in pet stores and pet supply shops as well as in veterinarian clinics. The most common pet species are cockatiels and parakeets, but many other species are now kept in captivity, in zoos or private collections or as pets. Keeping these birds healthy and in breeding condition requires specialised knowledge of diet and behavior especially if they are to be bred successfully. Captive breeding programs are now vital in the preservation of some avian species because of habitat destruction, and there are many facts about these birds that were learned from observations in captivity because it is so difficult to observe them in the wild.

Techniques

The tools and techniques of ornithology are varied and new inventions and approaches are quickly incorporated. The techniques may be broadly dealt under the categories of those that are applicable to specimens and those that are used in the field, however the classification is rough and many analysis techniques are usable both in the laboratory and field or may require a combination of field and laboratory techniques.

Collections

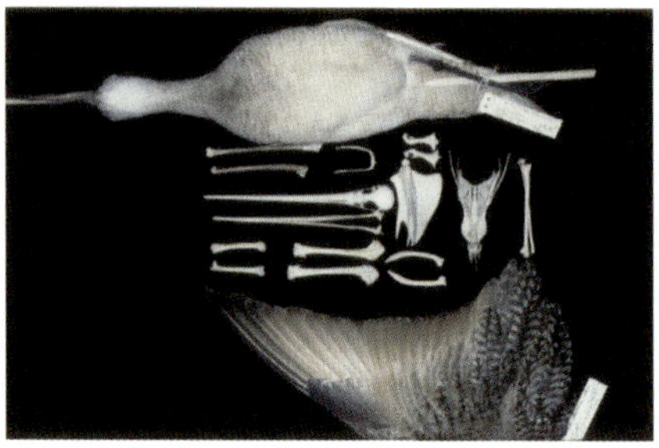

Bird preservation techniques

The earliest approaches to modern bird study involved the collection of eggs, a practice known as oology. While collecting became a pastime for many amateurs, the labels associated with these early egg collections made them unreliable for the serious study of bird breeding. In order to preserve eggs, a tiny hole was pierced and the contents extracted. This technique became standard with the invention of the blow drill around 1830. Egg collection is no longer popular; however historic museum collections have been of value in determining the effects of pesticides such as DDT on physiology. Museum bird collections continue to act as a resource for taxonomic studies.

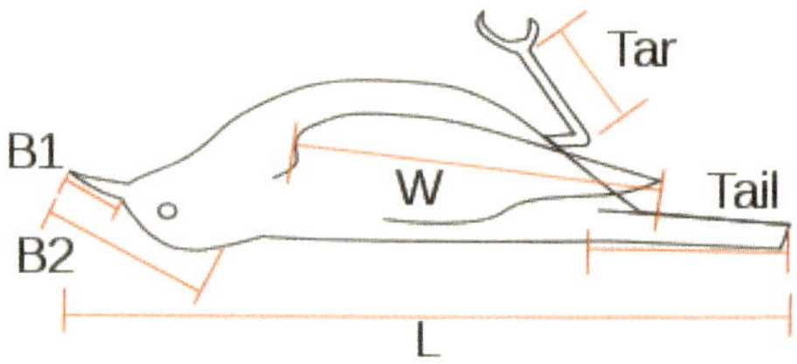

Morphometric measurements of birds are important in systematics

The use of bird skins to document species has been a standard part of systematic ornithology. Bird skins are prepared by retaining the key bones of the wings, leg and skull along with the skin and feathers. In the past, they were treated with arsenic to prevent fungal and insect (mostly dermestid) attack. Arsenic, being toxic, was replaced by borax. Amateur and professional collectors became familiar with these skinning techniques and started sending in their skins to museums, some of them from distant locations. This led to the formation of huge collections of bird skins in museums in Europe and North America. Many private collections were also formed. These became references for comparison of species and the ornithologists at these museums were able to compare species from different locations, often places that they themselves never visited. Morphometrics of these skins, particularly the lengths of the tarsus, bill, tail and wing became important in the descriptions of bird species. These skin collections have been utilized in more recent times for studies on molecular phylogenetics by the extraction of ancient DNA. The importance of type specimens in the description of species make skin collections a vital resource for systematic ornithology. However, with the rise of molecular techniques, it has now become possible to establish the taxonomic status of new discoveries, such as the Bulo Burti boubou (*Laniarius liberatus*, no longer a valid species) and the Bugun liocichla (*Liocichla bugunorum*), using blood, DNA and feather samples as the holotype material.

Other methods of preservation include the storage of specimens in spirit. Such wet-specimens have special value in physiological and anatomical study, apart from providing better quality of DNA for molecular studies. Freeze drying of specimens is another technique that has the advantage of preserving stomach contents and anatomy, although it tends to shrink making it less reliable for morphometrics.

In the Field

The study of birds in the field was helped enormously by improvements in optics. Photography made it possible to document birds in the field with great accuracy. High power spotting scopes today allow observers to detect minute morphological differences that were earlier possible only by examination of the specimen *in the hand*.

A bird caught in a mist net

The capture and marking of birds enables detailed studies of life-history. Techniques for capturing birds are varied and include the use of bird liming for perching birds, mist nets for woodland birds, cannon netting for open area flocking birds, the bal-chatri trap for raptors, decoys and funnel traps for water birds.

A researcher measures a wild woodpecker. The bird's right leg has a metal identification tag.

The bird in the hand may be examined and measurements can be made including standard lengths and weight. Feather moult and skull ossification provide indications of age and health. Sex can be determined by examination of anatomy in some sexually non-dimorphic species. Blood samples may be drawn to determine hormonal conditions in studies of physiology, identify DNA markers for studying genetics and kinship in studies of breeding biology and phylogeography. Blood may also be used to pathogens and arthropod borne viruses. Ectoparasites may be collected for studies of coevolution and zoonoses. In many of cryptic species, measurements (such as the relative lengths of wing feathers in warblers) are vital in establishing identity.

Captured birds are often marked for future recognition. Rings or bands provide long-lasting identification but require capture for the information on them to be read. Field identifiable marks such as coloured bands, wing tags or dyes enable short-term studies where individual identification is required. Mark and recapture techniques make demographic studies possible. Ringing has traditionally been used in the study of migration. In recent times satellite transmitters provide the ability to track migrating birds in near real-time.

A California condor marked with wing tags

Techniques for estimating population density include point counts, transects and territory mapping. Observations are made in the field using carefully designed protocols and the data may be analysed to estimate bird diversity, relative abundance or absolute population densities. These methods may be used repeatedly over large time spans to monitor changes in the environment. Camera traps have been found to be a useful tool for the detection and documentation of elusive species, nest predators and in the quantitative analysis of frugivory, seed dispersal and behaviour.

In the Laboratory

Many aspects of bird biology are difficult to study in the field. These include the study of behavioural and physiological changes that require a long duration of access to the bird. Non-destructive samples of blood or feathers taken during field studies may be studied in the laboratory. For instance, the variation in the ratios of stable hydrogen isotopes across latitudes makes it possible to roughly establish the origins of migrant birds using mass spectrometric analysis of feather samples. These techniques can be used in combination with other techniques such as ringing.

The first attenuated vaccine developed by Louis Pasteur was for fowl cholera and was tested on poultry in 1878. Anti-malarials were tested on birds which harbour avian-malarias. Poultry continues to be used as a model for many studies in non-mammalian immunology.

Studies in bird behaviour include the use of tamed and trained birds in captivity. Studies on bird intelligence and song learning have been largely laboratory based. Field researchers may make use of a wide range of techniques such as the use of dummy owls to elicit mobbing behaviour, dummy males or the use of call playback to elicit territorial behaviour and thereby to establish the boundaries of bird territories.

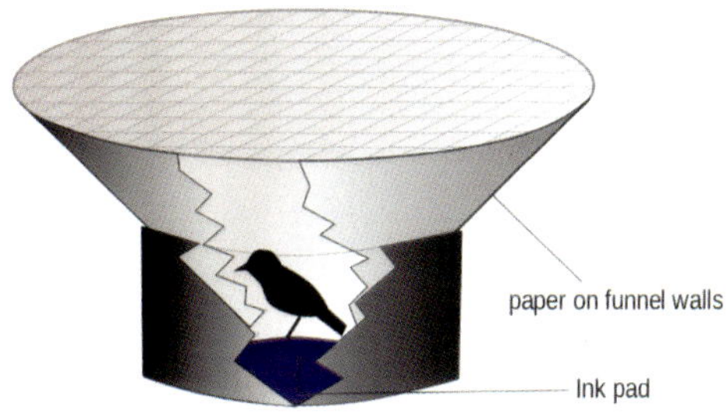

An Emlen funnel is used to study the orientation behaviour of migratory birds in a laboratory. Experimenters sometimes place the funnel inside a planetarium to study night migration.

Studies of bird migration including aspects of navigation, orientation and physiology are often studied using captive birds in special cages that record their activities. The Emlen funnel for instance makes use of a cage with an inkpad at the centre and a conical floor where the ink marks can be counted to identify the direction in which the bird attempts to fly. The funnel can have a transparent top and visible cues such as the direction of sunlight may be controlled using mirrors or the positions of the stars simulated in a planetarium.

The entire genome of the domestic fowl (*Gallus gallus*) was sequenced in 2004 and was followed in 2008 by the genome of the zebra finch (*Taeniopygia guttata*). Such whole genome sequencing projects allow for studies on evolutionary processes involved in speciation. Associations between the expression of genes and behaviour may be studied using candidate genes. Variations in the exploratory behaviour of great tits (*Parus major*) have been found to be linked with a gene orthologous to the human gene *DRD4* (Dopamine receptor D4) which is known to be associated with novelty-seeking behaviour. The role of gene expression in developmental differences and morphological variations have been studied in Darwin's finches. The difference in the expression of *Bmp4* have been shown to be associated with changes in the growth and shape of the beak.

The chicken has long been a model organism for studying vertebrate developmental biology. As the embryo is readily accessible, its development can be easily followed (unlike mice). This also allows the use of electroporation for studying the effect of adding or silencing a gene. Other tools for perturbing their genetic makeup are chicken embryonic stem cells and viral vectors.

Collaborative Studies

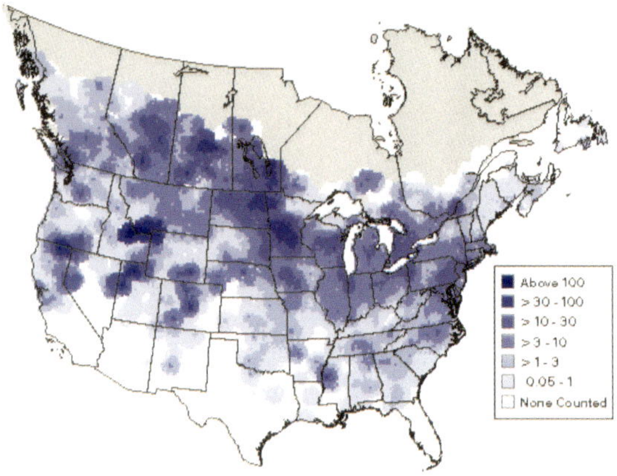

Summer distribution and abundance of Canada goose using data from the North American Breeding Bird Surveys 1994–2003

With the widespread interest in birds, it has been possible to use a large number of people to work on collaborative ornithological projects that cover large geographic scales. These citizen science projects include nationwide projects such as the Christmas Bird Count, Backyard Bird Count, the North American Breeding Bird Survey, the Canadian EPOQ or regional projects such as the Asian Waterfowl Census and Spring Alive in Europe. These projects help to identify distributions of birds, their population densities and changes over time, arrival and departure dates of migration, breeding seasonality and even population genetics. The results of many of these projects are published as bird atlases. Studies of migration using bird ringing or colour marking often involve the cooperation of people and organizations in different countries.

Applications

Wild birds impact many human activities while domesticated birds are important sources of eggs, meat, feathers and other products. Applied and economic ornithology aim to reduce the ill effects of problem birds and enhance gains from beneficial species.

Red-billed quelea are a major agricultural pest in parts of Africa.

The role of some species of birds as pests has been well known, particularly in agriculture. Granivorous birds such as the queleas in Africa are among the most numerous birds in the world and foraging flocks can cause devastation. Many insectivorous birds are also noted as beneficial in agriculture. Many early studies on the benefits or damages caused by birds in fields were made by analysis of stomach contents and observation of feeding behaviour. Modern studies aimed to manage birds in agriculture make use of a wide range of principles from ecology. Intensive aquaculture has brought humans in conflict with fish-eating birds such as cormorants.

Large flocks of pigeons and starlings in cities are often considered as a nuisance and techniques to reduce their populations or their impacts are constantly innovated. Birds are also of medical importance and their role as carriers of human diseases such as Japanese Encephalitis, West Nile Virus and H5N1 have been widely recognised. Bird strikes and the damage they cause in aviation are of particularly great importance, due to the fatal consequences and the level of economic losses caused. It has been estimated that the airline industry incurs worldwide damages of US$1.2 billion each year.

Many species of birds have been driven to extinction by human activities. Being conspicuous elements of the ecosystem, they have been considered as indicators of ecological health. They have also helped in gathering support for habitat conservation. Bird conservation requires specialized knowledge in aspects of biology, ecology and may require the use of very location specific approaches. Ornithologists contribute to conservation biology by studying the ecology of birds in the wild and identifying the key threats and ways of enhancing the survival of species. Critically endangered species such as the California condor have had to be captured and bred in captivity. Such ex-situ conservation measures may be followed by re-introduction of the species into the wild.

Birds

Birds (Aves) are a group of vertebrates which evolved from dinosaurs. They are endothermic, with feathers.

Modern birds are toothless: they have beaked jaws. They lay hard-shelled eggs. They have a high metabolic rate, a four-chambered heart, and a strong yet lightweight skeleton.

Birds live all over the world. They range in size from the 5 cm (2 in) bee hummingbird to the 2.70m (9 ft) ostrich. They are the class of tetrapods with the most living species: about ten thousand. More than half of these are passerines, sometimes known as perching birds.

Birds are the closest living relatives of the Crocodilia. The fossil record shows that birds evolved from feathered theropod dinosaurs. Modern birds are not descended from *Archaeopteryx*. They first appeared during the Cretaceous , about 100 million years ago.

Birds diversified dramatically round about the time of the Cretaceous–Palaeogene extinction event 66 million years ago, which killed off all the non-avian dinosaur lines. Birds, especially those in the southern continents, survived this event and then migrated to other parts of the world.

Primitive bird-like dinosaurs are in the broader group Avialae. They have been found back to the mid-Jurassic period, around 170 million years ago. Many of these early “stem-birds”, such as *Anchiornis*, were not yet capable of fully powered flight. Many had primitive characteristics like teeth in their jaws, and long bony tails.

Birds have wings which are more or less developed depending on the species. The only known groups without wings are the extinct moa and elephant birds. Wings, which evolved from forelimbs, gave birds the ability to fly. Later many groups evolved with reduced wings, such as ratites, penguins, and many island species of birds. The digestive and respiratory systems of birds are also adapted for flight. Some bird species in aquatic environments, particularly seabirds and some waterbirds, have evolved as good swimmers.

Some birds, especially crows and parrots, are among the most intelligent animals. Several bird species make and use tools. Many social species pass on knowledge across generations, a form of culture. Many species annually migrate great distances. Birds are social. They communicate with visual signals, calls, and bird songs. They have social behaviours such as cooperative breeding and hunting, flocking, and mobbing of predators.

Most bird species are socially monogamous, usually for one breeding season at a time, sometimes for years, but rarely for life. Other species are polygynous (one male with many females) or, rarely, polyandrous (one female with many males). Birds produce offspring by laying eggs which are fertilised by sexual reproduction. They are often laid in a nest and incubated by the parents. Most birds have an extended period of parental care after hatching. Some birds, such as hens, lay eggs even when not fertilised, though unfertilised eggs do not produce offspring.

Many species of birds are eaten by humans. Domesticated and undomesticated birds (poultry and game) are sources of eggs, meat, and feathers. Songbirds, parrots, and other species are popular as pets. Guano is harvested for use as a fertiliser. Birds figure throughout human culture. About 120–130 species have become extinct due to human activity since the 17th century, and hundreds more before then. Human activity threatens about 1,200 bird species with extinction, though efforts are underway to protect them. Recreational birdwatching is an important part of the ecotourism industry.

Bird Colours

Birds come in a huge range of colours. These colours can be useful to a bird in two ways. Camouflage colours help to hide the bird, and bright colours identify the bird to others of the same species. Often the male is brightly coloured while the female is camouflaged.

Bird Camouflage

Grey partridge

This Scops owl is almost invisible against the tree

Many birds are brown, green or grey. These colours make a bird harder to see; they camouflage the bird. Brown is the most common colour. Brown birds include sparrows, emus, thrushes, larks, eagles, falcons and the female birds of many species such as wrens, ducks, blackbirds and peacocks. When a brown bird is in long grass or among tree trunks or rocks, it is *camouflaged*. Birds that live in long grass often have brown feathers streaked with black which looks like shadows. A bittern is almost invisible in long reeds. Other birds, including starlings and minahs, are quite dark in colour, but are flecked with little spots that look like raindrops on leaves.

Many birds from hot countries are green or have some green feathers, particularly parrots. Birds that live in green trees often have green backs, even if they have bright-coloured breasts. From the back, the birds are *camouflaged*. This is very useful when sitting on a nest. The bird's bright-coloured breast is hidden. Budgerigars are bred in different colours such as blue, white and mauve, but in the wild, they are nearly all green and yellow. Even though they fly very well, they normally spend a lot of time on the ground, eating grass seeds. Their yellow and black striped back helps to hide them in the shadows made by long dry grass, while their green breasts are a similar colour to the leaves of gum trees.

Grey birds include most pigeons and doves, cranes, storks and herons. Grey birds are often rock-living birds like pigeons, or birds that sit on dead tree trunks looking like a broken branch. Water birds like herons often have a pale grey colour which makes it harder for a fish to notice that the bird is standing, looking down for something to catch. Water birds, no matter what colour they are on top, are often white underneath, so that when a fish looks up, the bird looks like part of the sky.

Black birds include crows, ravens and male blackbirds. Some birds that are dark colours spend quite a lot of time on the ground, hopping around in the shadows under bushes. Among these birds are the male blackbird and the Satin Bowerbird which is not black but very dark blue. Crows and ravens often perch high on bare trees in the winter, where their black shape against the sky looks like the dark bare branches.

Noticeable Colours

Many birds are not camouflaged, but stand out with vivid colours. They are usually male birds whose females are dull and camouflaged. The function of the colours is two-fold. First, the colours help them get mates, and second, the colours identify them to other males of the same species. Many birds are territorial, especially in the nesting season. They give out territory sounds and are easily seen. This lets other males know they will defend their territory. It sends out a "look elsewhere" signal to their competitors.

Pair of mandarin ducks. A good example of a general rule: male birds are the showy ones, females are dowdy

Some birds are famous for their colour and are named for it, such as the bluebird, the azure kingfisher, the golden pheasant, the scarlet macaw, the violet wren and the robin.

Many other birds are very brightly coloured, in countless combinations. Some of the most colourful birds are quite common, like pheasants, peacocks, domestic fowl and parrots. Colourful small birds include blue tits, the gold finches, humming birds, fairy wrens and bee eaters (which are also called *rainbow birds*). Some birds, like those of the bird of paradise in Papua New Guinea have such beautiful feathers that they have been hunted for them.

The peafowl is the best example of a display of colour to attract a mate. Also the male domestic fowl and junglefowl have long shiny feathers above his tail and also long neck feathers that may be a different colour to his wings and body. There are only a very few types of birds (like the eclectus parrot) where the female is more colourful than the male.

'Pied birds' are black and white. Black and white birds include magpies, pied geese, pelicans, and Australian magpies (which are not really magpies at all). Pied birds often have brightly coloured beaks and legs of yellow or red. The silver pheasant, with its long white tail striped with fine bars of black, has a brightly coloured face.

Common shelduck

Kingfisher

Flamingo

Golden oriole.

Himalayan bluetail

Flight

Most birds can fly. They do this by pushing through the air with their wings. The curved surfaces of the wings cause air currents (wind) which lift the bird. Flapping keeps the air current moving to create lift and also moves the bird forward.

Some birds can *glide* on air currents without flapping. Many birds use this method when they are about to land. Some birds can also *hover* and remain in one place. This method is used by birds of prey such as falcons that are looking for something to eat. Seagulls are also good at *hovering*, particularly if there is a strong breeze. The most expert *hovering* birds are tiny hummingbirds which can beat their wings both backwards and forwards and can stay quite still in the air while they dip their long beaks into flowers to feed on the sweet nectar.

This osprey at Kennedy Space Centre is hovering.

The soft feathers of an owl allow it to fly quietly.

Some birds, such as the quail, live mainly on the ground.

A cassowary cannot fly but can defend itself.

Penguin's flippers are good for swimming.

Types of Flight

Different types of birds have different needs. Their wings are adapted to suit the way they fly.

Large birds of prey, such as eagles, that spend a lot of time *soaring* on the wind have wings that are large and broad. The main flight feathers are long and wide. They help the eagle to stay on rising *air currents* without using much energy, while the eagle looks at the ground below, to find the next meal. When the eagle sees some small creature move, it can close its wings and fall from the sky like a missile, opening its great wings again to slow down as it comes to land. The world's largest eagle, the Philippine eagle has a wingspan of about 2 metres (6.7 ft) wide.

Birds that live in grassland areas or open forests and feed on fruit, insects and reptiles often spend a lot of time flying short journeys looking for food and water. They have wings that are shaped in a similar way to eagles, but rounder and not as good for soaring. These include many Australian birds like cockatoos.

Birds, such as geese, that migrate from one country to another fly very long distances. Their wings are big and strong, because the birds are large and they stock up on food for the long flight. Migrating water birds usually form family groups of 12-30 birds. They fly very high, making use of long streams of air that blow from north to south in different seasons. They are very well organised, often flying in a V pattern. The geese at the back do not have to flap so hard; they are pulled on by the wind of the ones at the front. Every so often, they change the leader so that the front bird, who does most work and sets the pace, can have a rest. Geese and swans are the highest-flying birds, reaching 8,000 metres or more when on migration. Geese often *honk* loudly while they are flying. It is thought that they do this to support the leader and help the young ones.

Birds that fly very quickly, such as swifts and swallows, have long narrow pointed wings. These birds need great speed because they eat insects, catching most of them while they are flying. These birds also migrate. They often collect in huge flocks of thousands of birds that move together like a whirling cloud.

Birds that live in bushes and branches have triangular wings that help the bird change direction. Many forest birds are expert at getting up speed by flapping and then gliding steadily among the trees, tilting to avoid things as they go. Members of the kingfisher family are expert at this type of flying.

Birds such as owls that hunt at night have wings with soft rounded feathers so that they do not flap loudly. Birds that are awake at night are called *nocturnal* birds. Birds that are awake during the day are *diurnal.*

A wandering albatross and Arctic tern might spend several years without coming to land. They can sleep while gliding and have wings which, when they are stretched right out, look like the wings of a jet plane.

Bird like chickens that feed mainly on the ground and only use their wings to fly to safety have small wings.

Flocks

Flocks of birds can be very highly organised in a way that takes care of all the flock members. Studies of small flocking birds like tree sparrows show that they clearly communicate with each other, as sometimes thousands of birds may fly in close formation and spiral patterns without *colliding* (or flying into each other).

Two common behaviours in flocking birds are *guarding* and *reconnaissance.* When a flock of birds is feeding it is common for one bird to perch on a high place to keep *guard* over the flock. In the same way, when a flock is asleep, often, one bird will remain awake. It is also common for large flocks to send one or two birds ahead of them when they are flying to a new area. The look-out birds can spy the *lie of the land* to find food, water and good places to perch.

Flightless Birds

Some birds do not fly. These include running birds like ostriches and emus and ocean-living birds, the large penguin family.

Ostriches and emus do not need to fly because although they feed and nest on the ground, their great size and their speed is their protection. Some other ground-feeding birds have not been so lucky. Some birds such as the dodo and the kiwi were ground-feeding birds that lived in safety on islands where there was nothing dangerous to eat them. They lost the power of flight. Kiwis are endangered because European settlement to New Zealand brought animals like cats, dogs and rats which kill kiwis and eat their eggs. However, kiwis and also the rare New Zealand ground parrot have survived. In the case of dodos, they were fat and delicious. They were killed and eaten by sailors until there was none left. Other flightless birds which have disappeared are the great auk and the moa.

Penguins spend a great deal of time at sea, where they are in danger from seals. On land, they usually live in areas where there were few dangers, until the arrival of European settlers with dogs and cats. Their wings have adapted to life in the sea and have become flippers which help them in swimming very fast.

Reproduction

Swans mate for life

Mating

Although birds are warm-blooded creatures like mammals, they do not give birth to live babies. They lay eggs like lizards do. Unlike most reptiles, the shell of a bird's egg is hard. The baby bird grows inside the egg and after a few weeks, breaks out, or *hatches*.

Birds in cold climates usually have a *breeding season* once a year in the spring. Migratory birds can have two springs and two mating seasons in a year. So can birds that live in hot climates.

When the breeding season arrives, the birds choose partners. Some birds are mated for life, like married couples. These birds include pigeons, geese, and cranes. Other birds look for new partners each year and sometimes a male bird or *cock* will have several wives.

For birds that choose new mates, part of the breeding season is *display*. The male bird will do all sorts of things to attract females. These include singing, dancing, showing off the feathers and building a beautiful nest. Some male birds have splendid feathers for attracting females. The most famous is the peacock who can spread the feathers above his tail into a huge fan.

A peacock display.

Emu nest.

A nest of house sparrows.

Nesting

Once the birds have found partners, they find a suitable place to lay eggs. The idea of what is a suitable place differs between species, but most build bird nests. Robins will make a beautiful little round nest of woven grass and carefully line it with feathers, bits of fluff and other soft things. Swallows like to nest near other swallows. They make nests from little blobs of clay, often on a beam near the roof of a building where it is well sheltered. Many birds like a hollow tree to nest in. Eagle's nests are often just piles of dead wood on the top of the tallest tree or mountain. Scrub turkeys scratch together a huge pile of leaves that may be 10 metres across. Guillemots lay their eggs on rock shelves with no nest at all. Their eggs are shaped so that they roll around in circles and do not fall off cliffs. A cuckoo does not make its own nest. It lays its egg in the nest of another bird and leaves it for them to care for. The cuckoo eggs are camouflaged to look like the host's eggs.

When the nest has been prepared, the birds mate so that the eggs are fertilised and the chicks will start growing. Unlike mammals, birds only have one opening as the exit hole for body fluids, and for reproduction. The opening is called the cloaca. A female bird, called a *hen*, has two ovaries, of which the left one usually produces eggs.

Most male birds have no sex organs that can be seen. But inside the male are two testes which produce sperm which is stored in the cloaca. Birds mate by rubbing their cloacas together, although with some birds, particularly large water birds, the male has a sort of a penis inside the cloaca.

Hatching

Once the *hen* has mated, she produces fertile eggs which have chicks growing inside them. She lays the eggs in the nest. There might be just one egg or a number of them, called a *clutch*. Emus might lay as many as fifteen huge dark green eggs in a clutch. After the eggs are laid, they are *incubated*, or kept warm so the chicks form inside. Most birds stay together for the whole nesting season, and one advantage is that the work is shared. Many birds take turns sitting on the eggs, so that each adult can feed.

This is not always the case. With emus, the male does all the *sitting* and all the baby-minding. With emperor penguins it is also the male that cares for the egg. There is only one egg, which he keeps on his feet and under his feathers, standing in a big group of males without feeding until the chick is hatched. While the eggs are hatching, the females are at sea, feeding, so that they can care for the chicks when they return.

Some birds put the eggs inside or on top of the mound of leaves and twigs. The mound acts like a compost heap. The decomposition of the rotting leaves causes the temperature to rise. This is heat released by the chemical action of bacterial and fungal respiration. It is the same reaction as that which keeps mammals and birds at a high temperature. The parents leave the mound. When the chicks hatch, they are able to feed themselves.

Many small birds take 2–4 weeks to hatch eggs. Albatrosses take 80 days. During this time the female loses a lot of her body weight.

The quickest hatching time is for the cuckoo. Some types of cuckoos take only 10 days. This means that when they hatch in the nest of their 'foster parents', the eggs that the parents have laid are not yet ready. Newborn cuckoos are naked, blind and ugly, but they are strong. They get under any eggs that are in the nest and throw them out before they hatch. That means that the cuckoo has the whole care of both parents. Baby cuckoos grow fast and often get bigger than the parents who feed them.

When baby birds hatch, in most types of birds, they are fed by both parents, and sometimes by older *aunties* as well. Their mouths are open all the time and are often very brightly coloured, which acts as a 'releaser', a trigger which stimulates the parent to feed them. For birds that eat grain and fruit, the parents eat and partly *digest* the food for the babies. It is then vomitted carefully into the baby's mouth.

A black redstart feeding chicks

Black swan and cygnets

A reed warbler feeding a baby cuckoo

Two sulphur crested cockatoos from a big flock are on the lookout

Families

Many birds, particularly those that mate for life, are very sociable and keep together in a family group which might be anything from 4 or 6 adult birds and their young to a very large flock.

As chicks grow they change the fluffy *down* that covers them as babies for real feathers. At this stage they are called *fledglings*. Other family members may help care for fledgling chicks, feeding them, and protecting them from attack while parents are feeding. When the *fledglings* have their new feathers, they come out of the nest to learn to fly. In some types of birds, like pigeons, the parents watch over this and as the young ones get stronger, will give them flying lessons, teaching them how to glide, how to fly in spirals and how to land like an expert.

Communication

Most birds are social animals, at least part of the time. They communicate to each other using sounds and displays.

Almost all birds make sounds to communicate. The types of noises that vary greatly. Some birds can *sing*, and they are called *songbirds*. Examples are robins, larks, canaries, thrushes, nightingales and crows. Birds that are *not* songbirds include pigeons, seagulls, eagles, owls and ducks. Parrots are not songbirds, even though they can be taught to sing human songs.

A favorite songbird, the European robin.

The crow of the rooster is a familiar bird call.

The pied currawong, an outstanding singer.

The jackdaws helped Lorenz to understand bird communication.

Songbirds

All birds make noises ('bird vocalisation'), but not all sing. Songbirds are passerines, many of which have beautiful melodic songs. Songs have different functions. Danger cries are different from territorial songs, and mating calls are a third type. Fledgling may also have different calls from adults. Recognition calls for partners are quite common.

As to where the song comes from, there are three kinds of species:

1. Those where the song is entirely inherited, and the bird always sings the same song in the same situations.

2. Those where the song is partly inherited, but the bird tunes it in by copying others. In this case the slight differences between the calls of different birds may be used by partners for identification.

3. Those where the song is entirely learnt, and the bird often copies sounds from its environment.

Most singing birds that are kept as pets, like canaries, have several tunes and some variations.

The same species of bird will sing different songs in different regions. A good example of this is the currawong. This is an Australia bird which is like a black and white crow. In the autumn, families get together in large flocks and do a lot of singing. Currawongs from some areas sing much more complex songs than others. Generally, currawongs from the Blue Mountains are the finest singers. The song of the currawong can be sung as a solo, but is often performed as a choir. One bird will take the lead and sing "Warble-warble-warble-warble!" All the other birds will join in and sing "Wooooooo!" When all the birds know the song, the choir will sing the "Warble" part and the soloist will sing the "Woo!". The song changes from year to year and from place to place.

Lorenz's Studies

The Austrian naturalist Konrad Lorenz studied the way in which birds communicate, or talk to each other. He found that each type of bird had a number of sounds which they made automatically, when ever they felt a certain way. Every sound had an action that went with it. So, if the bird was frightened, it acted frightened and made a frightened sound. This told the other birds around it that something frightening was happening.

If a flock of birds were flying over a field, they would be calling "Fly! Fly!" But a hungry bird, seeing something good to eat down below might start calling "Food! Food!" If other birds were also hungry, they would make the same call until more birds were calling "Food! Food!" than "Fly! Fly!". At this point, the mind of the flock would be changed. Some of the birds would start to yell "Fly downwards! Fly downwards!" as they sank from the sky, until the whole flock was all noisily calling the same thing.

These communication sounds are often short hard sounds like chirps, squeaks, squawks. and twitters. Sometimes the calls are longer and more musical. They include the "Rookety-coo" sound of a pigeon and the "Cockadoodledoo!" of a rooster. The bird cannot change these sounds. They always make them in the same way. The bird is locked into making each sound every time a particular idea comes into its head. The connection between how they feel and how they call is innate: they are born with it. Some calls in some species are learnt. Then, it is the tendency to learn which is inherited.

The Jackdaw of Altenberg

Konrad Lorenz noticed that when birds sing, they often use a lot of their regular calls as part of the song. Lorenz had a flock of jackdaws which were scattered during World War II. One day, an old bird returned. For many months she sat on the chimney singing her song, but in the song she kept making the call which Lorenz knew meant "Come home! Come home!" One day, to the great surprise of Lorenz, a male bird flew from a passing flock and joined her on the chimney. Lorenz was sure that it was her long-lost "husband" who had found his way home at last.

Evolution and Taxonomy

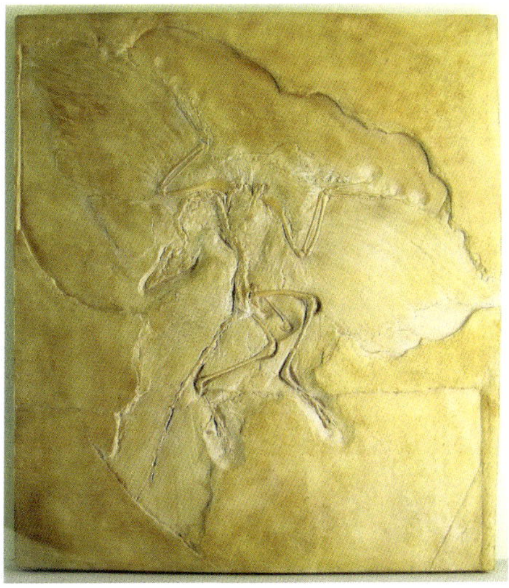

Archaeopteryx, the earliest known bird

Confuciusornis, a Cretaceous bird from China

Palaeontologists have found some exceptional places (lagerstätten) where fossils of early birds are found. The preservation is so good that on the best examples impressions of their feathers can be seen, and sometimes even the remains of meals they have eaten. From these remains we know that birds evolved from small carnivorous dinosaurs (theropods) in the Jurassic period. They radiated into a huge variety in the Lower Cretaceous. At the same time, their direct competitors, the pterosaurs, dwindled in numbers and variety, and became extinct at the end of the Mesozoic.

Birds are classified by taxonomists as 'Aves' (Avialae). Birds are the only living descendants of dinosaurs (strictly speaking, they *are* dinosaurs). Birds and Crocodilia are the only living members of the once-dominant Archosaur reptiles.

The class Aves is now defined as all the descendants of the most recent common ancestor of modern birds and *Archaeopteryx lithographica*.

The First Bird-like Creatures

Archaeopteryx, from the Upper Jurassic (some 150–145 million years ago), is the earliest bird which could fly. It is famous, because it was one of the first important fossils found after Charles Darwin published his ideas about evolution in the 19th century. By modern standards, *Archaeopteryx* could not fly very well. Other early fossil birds are, for example, *Confuciusornis*, *Anchiornis huxlei*, and other Paraves.

Many fossils of early birds and small dinosaurs have been discovered in the Liaoning Province of Northeast China. The fossils show that most small theropod dinosaurs had feathers. These deposits have preserved them so well that the impressions of their feathers can be clearly seen. This leads us to think that feathers evolved first as heat insulation, and only later for flight. The origin of birds lies in these small feathered dinosaurs.

Palaeontologists now agree that birds evolved from Maniraptora group of dinosaurs. This explains why one might say birds are living dinosaurs.

- Origin of birds
- Paraves

Birds and People

Canaries are often kept as pets for their beautiful songs.

The African grey parrot is a renowned talker.

Blue-winged teal. Ducks used to be shot for sport.

In many countries storks are thought to bring good luck.

Some birds are eaten as food. Most usually it is the chicken and its eggs, but people often also eat geese, pheasants, turkeys and ducks. Other birds are sometimes eaten are emus, ostriches, pigeons, grouse, quails, doves, woodcocks, and even songbirds. Some species have died out because they have been hunted for food, for example the dodo, and the passenger pigeon.

Many species have learned how to get food from people. The number of birds of these species has grown because of it. Seagulls and crows find food from garbage dumps. The feral pigeon (*Columba livia*), sparrows (*Passer domesticus* and starlings (*Sturnus vulgaris*) live in large numbers in towns and cities all over the world.

Sometimes people also use working birds. For example, homing pigeons carry messages. Nowadays people sometimes race them for sport. People also use falcons for hunting, and cormorants for fishing. In the past, people in mines often used a canary to see if there were bad gas methane in the air.

People often have colorful birds such as parrots and mynahs as pets. These intelligent birds are popular because they can copy human talking. Because of this, some people trap birds and take them to other countries to sell. This is not usually allowed these days. Most pet birds are specially bred and are sold in pet shops.

People can catch some bird diseases, for example *psittacosis*, *salmonellosis*, *campylobacteriosis*, Newcastle's disease, *mycobacteriosis*, influenza, *giardiasis*, and *cryptosporiadiosis*. In 2005, there was an epidemic of bird influenza spreading through some parts of the world, often called avian flu.

Some people have birdboxes in their gardens to give birds a place to nest and bird tables where birds can get food and water in very cold or very dry weather. This lets people see some small birds close up which are normally hidden away in bushes and trees.

Herpetology

Herpetology is the branch of zoology dealing with the study of reptiles and amphibians.

The study of reptiles and amphibians reveals humanity's strong and lasting curiosity about nature. Since the dawn of humanity, humans have studied reptiles and amphibians, if only as a source of food or because of the dangers of toxins or venoms. Herpetology not only offers many practical benefits—medicinal applications, conservation, environmental indicators, model organisms, and so forth—but also addresses people's desire to better understand the world around them.

Overview

Herpetology is concerned with poikilothermic, or cold-blooded, tetrapods (four-legged vertebrates). "Herps" include reptiles like lizards, snakes, crocodiles, and turtles, and amphibians like frogs, toads, salamanders, newts, waterdogs, mudpuppies, sirens, and caecilians. Herpetology is concerned with the study of such extant taxa and even extinct taxa.

Reptiles and amphibians share poikilothermy, and sometimes a superficial resemblance to one another (e.g. salamanders and lizards are often confused), but these two groups are quite distinct. The essential difference is that amphibians include all tetrapods that do not have amniotic eggs, while reptiles are tetrapods and amniotes (animals whose embryos are surrounded by an amniotic membrane that encases it in amniotic fluid). There are many other differences. For example, amphibians tend to have permeable skin that allows for the exchange of gases, are often bound to water for at least part of their lives, have glandular skin, and many of them produce toxic secretions in some of their skin glands, warding off predators and generally tasting bad. Reptiles, by contrast, typically have a dry watertight skin, usually protected by scales, that normally has few if any glands. Many species of reptiles spend no part of their lives anywhere near water, and they all have lungs. Living reptilians usually, if not always, lay eggs or give birth on land, including marine turtles, which only come ashore for this purpose. Again, extinct creatures may have exhibited some differences.

Despite these differences between amphibians and reptiles, they form a common focus under the umbrella of herpetology. Furthermore, despite the differences between these two groups and fish (non-tetrapod vertebrates), it also is not uncommon for herpetological and ichthyological (study of fish) scientific societies to "team up," publishing joint journals and holding joint conferences in order to foster the exchange of ideas between the fields One of the most prestigious herpetology societies is an example of this: the American Society for Ichthyology and Herpetology.

Importance

The interest in herpetology is ancient, with observations being made by Aristotle and even references in the first book of the Bible. Discussions of reptiles and amphibians can be found in early Egyptian religions, Shamanism, and religions in pre-Colombian Americas, where the Quetezecoatly was the serpent that was the law giver and culture bearer.

Because reptiles and amphibians offer important economic, ecological, aesthetic, and medicinal values, herpetology offers numerous benefits to humanity.

For example, amphibians are often very sensitive to environmental changes, offering a visible warning to humans that significant changes are taking place.

Some toxins and venoms produced by reptiles and amphibians are useful in human medicine. Currently, some snake venom has been used to create anti-coagulants that work to treat stroke victims and heart attack cases. Medicine produced from amphibians now are used to treat heart aliments, bacterial infections, skin and colon cancers, depression, and chronic pain, among others. Amphibian toxins continue to be studied by herpetologists for possible applications. For example, a poison frog from South America *(Epipedobates tricolor)* secretes a non-addictive painkiller 200 times more potent than morphine, offering promise in this area if the toxicity can be neutralized.

The axolotl serves as model organism for research

Reptiles and amphibians are a critical element in the food chains of most ecosystems, and sometimes a keystone species whose removal can drastically alter the populations of other organisms. The consumption by reptiles and amphibians of rodents and insect pests aid in control of these animals, which can be serious agricultural pests; furthermore, reducing the numbers of insects can lower the prevalence of insect-borne diseases. Thus, studies aimed at maintaining the balance of nature are important. The skin of crocodilians, snakes, and lizards are used in leather goods, such as shoes, handbags, gloves, and belts, and thus study of these animals can offer direct economic applications.

Some reptiles likewise present threats to people, whether because they are venomous, like some snakes, or can attack humans, such as some crocodilians, or their skin harbors salmonella, a bacterial disease, and thus their study aids in preventing such dangers.

Some reptiles and amphibians are model organisms for research study. Axolotls, for example, are used extensively in scientific research into such areas as regeneration ability, heart defects, and vertebral development. Their ability to regenerate most body parts, ease of breeding, and large embryos are main reasons for their applicability as a model organism for research.

People with an avid interest in herpetology and who keep different reptiles or amphibians, often refer to themselves as "herpers."

Reptiles

Reptiles are vertebrate, or backboned, animals constituting the class Reptilia. They are characterized by a combination of features, none of which alone could separate all reptiles from all other animals. Among these features are

- cold-bloodedness;
- the presence of lungs;
- direct development, without larval forms as in amphibians;
- a dry skin with scales but not feathers (a characteristic of birds) or hair (a characteristic of mammals);
- an amniote egg;
- internal fertilization;
- a three- or four-chambered heart;
- two aortic arches (blood vessels) carrying blood from the heart to the body; mammals have only one aorta, the left; birds also have but one, the right;
- a metanephric kidney;
- twelve pairs of head (cranial) nerves; amphibians have ten; and
- skeletal features such as
    - limbs with usually five clawed fingers or toes,
    - at least two spinal bones (sacral vertebrae) associated with the pelvis; amphibians have but one,
    - a single ball-and-socket connection (condyle) at the head-neck joint instead of two, as in advanced amphibians and mammals, and
    - an incomplete or complete partition (the secondary palate) along the roof of the mouth, separating the food and air passageways so that breathing can continue while food is being chewed.

These and other traditional defining characteristics of reptiles have been subjected to considerable modification in recent times. The extinct flying reptiles, called pterosaurs, are now thought to have been warm-blooded and covered with hair; and the dinosaurs are also now considered by many authorities to have been warm-blooded. The earliest known bird, *Archaeopteryx*, is now regarded by many to have been a small dinosaur, despite its covering of feathers; and the extinct ancestors of the mammals, the therapsids, or mammallike reptiles, are also believed to have been warm-blooded and haired. Proposals have been made to reclassify the pterosaurs, dinosaurs, and certain other groups out of the class Reptilia into one or more classes of their own, and these issues are now receiving a great deal of attention from paleontologists and zoologists.

Cold-bloodedness

Reptiles are cold-blooded. That is, they lack the ability to regulate their metabolic heat (heat derived from the oxidation, or "burning," of food and from other processes) for the production of sustained body warmth and a constant body temperature. Cold-bloodedness, however, does not mean that a reptile is necessarily cold. A lizard basking in the Sun may have a higher body temperature than a mammal but must move into the shade to keep from overheating. Because cold-bloodedness is a misleading term, biologists employ two others instead, describing reptiles as poikilothermic and ectothermic. *Poikilothermy* refers to the condition in which body temperature varies with the temperature of the environment; it is contrasted with homeothermy, a characteristic of birds and mammals, in which body temperature remains essentially the same through a wide range of environmental temperatures. *Ectothermy* refers to the condition in which an animal depends on an external source, such as the Sun, rather than its own metabolism, for body warmth. Birds and mammals, which use their internal metabolic heat for body warmth, are referred to as endothermic.

Respiration

All reptiles possess lungs, and none passes through an aquatic larval stage with gills, as do many of the amphibians. In snakes, presumably as an adaptation to their long, thin bodies, the left lung is reduced in size or entirely lacking. Although lungs are the primary means of respiration in all reptiles and the only means of respiration in most reptiles, a number of species are also able to utilize other parts of the body for the absorption of oxygen and the elimination of carbon dioxide. In aquatic turtles, for example, the tissues (mucous membranes) lining the insides of the mouth are capable of extracting oxygen from the water; some file snakes, family Acrochordidae, and sea snakes, family Hydrophiidae, as well as the soft-shelled turtle, *Trionyx*, can use their skin for respiration when submerged.

Skin and Scales

Part of the ability of the amphibians'descendants, the reptiles, to invade dry-land environments was the development of a dry skin that served as a barrier to moisture and greatly reduced the loss of body water. The reptile skin, like that of other vertebrate animals, consists of two main parts: an outer epidermis and an underlying dermis. The epidermis produces horny, or keratinized (like fingernails), scales on its upper surface. These scales are not the same as (that is, not homologous to) the scales of fishes, which are bony, are formed in the dermis, and lie beneath the epidermis. The reptile's scales increase the skin's resistance to water, further reducing moisture loss; some scales may be modified for specialized functions, such as protective spines. Reptile scales may be small and overlapping, as in many lizards, or large and adjoining, as in turtles, where they are commonly called scutes. Some reptiles also have bony plates or nodules formed and lying within the dermis. Called dermal scales or osteoderms, these bony plates are similar in origin to fish scales. When present in lizards and crocodilians, the dermal scales are separated from one another, with each usually lying beneath and supporting an epidermal scale above. In turtles the bony plates are fused together to form a bony shell beneath the epidermal scales.

In lizards and snakes the scales do not increase in size as the animal grows; consequently, the old scales must be periodically shed and replaced by a new set of somewhat larger scales. Shedding may also occur when the outer layer becomes worn or when much food is consumed, as well as for causes not yet fully understood. In the shedding, or molting, process, also called ecdysis, the older upper layer of the epidermis with its attached scales loosens and breaks away from a newer layer that has developed beneath it. In turtles and crocodilians the large epidermal scales, or scutes, are not molted but are retained and are enlarged and thickened by additional layers of keratin from beneath; the uppermost layers of the scutes, however, may be lost through wear or other factors.

The Amniote Egg

A necessary part of the invasion of dry-land environments by the early reptiles was the development of an egg that could be laid out of water without drying up and that could “breathe” air rather than water. This egg, developed by the first reptiles, was the amniote egg, so named because it contains a membrane called the amnion. The amniote egg is found not only in reptiles but also in birds and (ancestral) mammals, and all three groups are sometimes collectively referred to as amniotes.

The amniote egg is different from the fishlike egg of most amphibians. It is enclosed in a protective shell, which is either flexible and leathery or rigid and calcareous. Within the shell is the fertilized egg cell lying on top of a large mass of yolk. The yolk is surrounded by a membrane (the yolk sac) and provides nourishment for the developing embryo. As the fertilized egg cell divides and redivides, and the embryo begins to form and grow, a folded membranous tissue grows up around the embryo, enclosing it in a double-walled sac. The outer wall of the sac is called the chorion, the inner wall, the amnion. The embryo is surrounded by fluid held with the amnion. The fluid provides the embryo with the aquatic environment it obviously still requires but which in amphibians is supplied by the waters of a pond or stream.

Another sac, called the allantois, projects from the embryo’s lower digestive tract (the hind gut) and acts as a bladder to receive the embryo’s waste products. The allantois sac becomes quite large, expanding out until its wall joins that of the chorion to form the chorioallantoic membrane, which is pressed up against the inside of the shell. Not only does the allantois serve as a bladder, receiving and storing insoluble wastes, but it also acts as a sort of “lung,” allowing oxygen and carbon dioxide to pass to and from the embryo through the slightly porous (permeable) egg shell.

Internal Fertilization

Because the egg cell reaches the outside environment surrounded by the shell of the egg, it is necessary that it be fertilized before it leaves the female’s body; thus, in all reptiles fertilization is internal, with males depositing sperm within the females’genital tracts. In snakes and lizards the male organ, called the hemipenes, is actually a pair of structures, with one of the pair, or hemipenis, situated internally on each side of the male's vent. Each hemipenis, which itself may be forked, is a functional structure: either one may be protruded from the vent and used in mating, the choice usually depending upon the placement of the male's mate. In turtles and crocodilians there is a single penis, which serves only for the transmission of sperm and not also for the elimination of excretory products, as in mammals. In the lizardlike tuatara, *Sphenodon*, the only living species in its order, the male lacks a copulatory organ, and mating is accomplished by the pressing together of the male's and female's cloacae, as in most birds.

The Heart

Except for crocodilians, which have a four-chambered heart, all reptiles have a three-chambered heart consisting of two atria and one ventricle. The chamber called the right atrium receives deoxygenated, or "spent," blood returning from the body tissues. It passes this blood into the ventricle, from where it is pumped to the lungs for oxygenation. The oxygenated blood from the lungs returns to the left atrium and once again enters the same ventricle, from which it is pumped to the body tissues.

Even in the three-chambered heart, however, as recent research has shown in contrast to earlier beliefs, there is little mixing of oxygenated and deoxygenated blood. This has been achieved by the development within the ventricle of interconnected "subchambers" within which a sequence of changes in blood pressure takes place. Several anatomical variations occur among the reptiles, but a simplified description of the lizard heart will serve to illustrate one way in which this circulatory efficiency was accomplished.

The ventricle of the lizard heart is incompletely partitioned into two subchambers by a muscular ridge that descends from the roof of the heart almost to the floor. The right subchamber is called the right ventricle, or cavum pulmonale; it leads to the lungs. The left subchamber is called the left ventricle, or cavum venosum; it receives blood from the right atrium and leads to the body. The two subchambers are connected not only beneath the partition but also across its incomplete rearward end.

A third subchamber, called the cavum arteriosum, is situated in the upper wall of the right ventricle; it receives blood from the left atrium and is connected through a valve-controlled opening to the left ventricle.

When the two atria contract, oxygenated blood from the left atrium enters the third subchamber, or cavum arteriosum, pressing against and shutting the valves controlling the opening into the left ventricle; this closes off the third subchamber and temporarily holds its contained blood in storage. At the same time the deoxygenated blood in the right atrium has been pumped into the left ventricle, filling it to overflowing, the excess blood moving across the open posterior end of the muscular partition into the right ventricle. The three ventricular subchambers are now filled with blood; oxygenated blood in the third subchamber (cavum arteriosum) and deoxygenated blood in the two ventricles.

Resistance to blood flow is lower in the pulmonary (heart-lung) circuit than in the systemic (heart-body) circuit, so that when the ventricles contract, the deoxygenated blood in the right ventricle follows the path of least resistance and proceeds through the pulmonary artery into the lungs. The emptying of the right ventricle causes more deoxygenated blood from the left ventricle to move around the open end of the partition into the right ventricle and to continue on to the lungs. The contraction of the ventricle also brings the partitioning muscular ridge into full contact with the floor of the heart, sealing off any flow from the right ventricle back into the left ventricle.

As the ventricles continue their contraction, the pressure on the blood held within the third subchamber (cavum arteriosum) exerts a reverse force on the valves, opening the passageway between the third subchamber and the left ventricle. The oxygenated blood in the third subchamber now moves into the left ventricle and from there into the aortas leading to the body circulation.

Metanephric Kidney

Adult amphibians have an opisthonephric kidney, that is, one that was developed from the main mass (except for the foremost end) of kidney-forming tissue in the embryo; the collecting tubules of the amphibian's opisthonephric kidney are connected to a drainage tube called the archinephric, or Wolffian, duct. In contrast, the metanephric kidney of adult reptiles (and birds and mammals) arises from the rearmost part of the kidney-forming tissue of the embryo — the front and middle portions having given rise to the pronephric and mesonephric kidneys, which eventually degenerate. The collecting tubules of the metanephric kidney are connected to a "newly evolved" drainage tube, the ureter. The Wolffian duct, having lost its excretory function, becomes partly involved in the transmission of sperm in males but is a degenerate vestige in females. In mammals the ureter drains into the bladder, which empties to the outside through another tube called the urethra. Crocodilians and snakes lack bladders, but even where one is present, as in most lizards and turtles, it is formed simply by an outpocketing of the cloaca and has no connection with the ureter. The cloaca, whose name comes from the Latin word for sewer, is a body chamber leading to the outside and into which empty the excretory products of the kidneys, the waste products of the intestines, and the reproductive products of the testes and ovaries. The reptilian ureter empties into the cloaca, and if a bladder is present it receives and holds the urine overflow.

Skull

The reptilian skull ranges from the reduced, loosely joined, or kinetic, skull of snakes to the large, solid skull of crocodilians. On each side of the skull behind the eye sockets there may be one or two openings formed at the junctions of certain bones. The openings are believed to be an adaptation for more-efficient functioning of the jaw muscles, possibly by allowing them to bulge outward when the jaws are closed. Because the positions of the openings, as determined by the bones surrounding them, were considered critical in the classification of reptiles, skull types were designated by the row, or arch, of bones beneath each opening rather than by the openings themselves. The Greek root *apse*, meaning "arch," was used in the naming of the skull types. A reptile skull without any temporal openings, and hence without arches, was termed *anapsid* (*an*, "without"; *apsid*, "arches"). Anapsid skulls are characteristic of the extinct, primitive cotylosaurs and of the living turtles.

When openings are present, their positions are best described in relation to the two skull bones extending from behind the eye socket to the rear of the head: the postorbital and squasmosal bones.

In the euryapsid-type skull there is a single opening high on each side of the head situated in the upper part of the junction between the postorbital and squamosal bones, which form the opening's lower border, or arch. In the parapsid-type skull, considered by some to be only a variant of the euryapsid, there is also a single opening high on each side of the head, but it is located above the postorbital and squamosal bones; its lower border, or arch, is formed by the supratemporal and postfrontal bones. An euryapsid-type skull was present in the extinct plesiosaurs, a parapsid-type skull in the extinct *Ichthyosaurus*.

A skull with a single low opening on each side of the head, situated in the lower part of the junction between the postorbital and squamosal bones and bordered below by parts of the squamosal, quadratojugal, and jugal bones, is termed *synapsid*. The extinct mammallike reptiles, order Therapsida, from which the mammals evolved, had synapsid skulls.

If a reptile skull has two openings, one high and one low, on each side of the head, situated in both the upper and lower parts of the junction between the postorbital and squamosal bones, it is classified as diapsid. Diapsid skulls are characteristic of dinosaurs and of the living crocodilians. Modern lizards and snakes also have diapsid skulls, but in the lizards the arch of the lower opening has been lost during evolution, leaving only the single upper opening and its arch. In snakes both the lower and upper arches have been lost, leaving no openings and very little bone. The reduction of bone and the increase of space in this area contribute to the movability of the snake's skull.

Sense Organs

Vision is the most commonly employed reptilian sense for the detection of objects in the environment. The most notable exceptions to this are the blind snakes, which burrow through the ground and have only vestigial eyes present beneath the scale on the head. Some turtles and lizards that are active by day — and perhaps some snakes, as well — definitely are able to see colors rather than simply shades of gray. Snakes and some lizards are noted for their unblinking stares, having apparently lost the ability to shut their eyes. Technically, the reverse might be said to be true. In these reptiles the lower eyelid has become transparent and is raised and permanently fused to the vestigial remnant of the upper lid. Thus, in a sense, their eyes are permanently closed.

The reptilian eye is permanently "set" for distance vision. Muscular action is required to look at nearer things. In all reptiles but snakes, near vision is achieved by contractions of muscles in the ciliary body surrounding the lens, forcing the lens into a more rounded shape. In snakes the contraction of special muscles forces the lens forward in adjustments for near vision.

A functioning so-called "third eye," or parietal body, is present in the upper skull of some lizards. It has a cornea, lens, and retina and is able to respond to differences in light and dark, but it cannot distinguish shapes. The parietal body is involved in the control of certain secretions and in certain adaptive behaviors. The lizardlike tuatara has a parietal body, as well.

The ear structure in most reptiles consists of an eardrum, which lies close to or near the surface of the skin (there is no external ear); a small bone, called the columella or stapes, which conducts the vibrations of the eardrum through the middle ear; and the inner ear, which receives the vibrations being transmitted by the columella. Turtles have well-developed ear structures but usually give little evidence of hearing ability. Certain tortoises (land turtles), however, appear sensitive to low sounds of about 100 Hz.

Snakes lack both the eardrum and the middle-ear cavity but retain the columella, which is positioned against the lower part of the skull at the jaw hinge (the quadrate bone). Lacking eardrums, a snake cannot "hear" airborne sounds in the usual sense. Experiments have shown, however, that the snake's inner ear does respond to low-frequency airborne sounds, or pressure vibrations, possibly by transmission of the vibrations through the skull bones. A snake can readily detect earthborne sounds, or vibrations.

The Jacobson's, or the vomeronasal, organ is a specialized chemical detector located in the roof of the reptilian mouth. It is present in rudimentary form in turtles and crocodilians but is highly developed in snakes and lizards. The rapid in-and-out flicking of a snake's tongue is to pick up tiny chemical particles, which adhere or dissolve in the moisture on the tongue and are then conveyed back to the two openings of the Jacobson's organ for identification. Pit vipers also possess organs that are highly sensitive to heat.

Origin and Evolution

The first known reptiles appeared during the Pennsylvanian Period (or upper Carboniferous Period), about 300 million years ago. The oldest of these, *Hylonomus*, was about 70 cm (27 in) long and lizardlike in shape. It was discovered in the Coal Measures of Nova Scotia. The reptiles evolved from amphibians known as anthracosaurs, which were part of the labyrinthodont group (amphibians with highly complex patterns in the surface of the teeth). The reptiles probably evolved during the Mississippian Period (or lower Carboniferous), about 40 million years earlier, and a fossil found in Scotland in the late 1980s, buried in the mudstone from an ancient lake bed, appears to support this view. Reptiles such as *Seymouria*, from the Permian Period about 275 million years ago, retain signs of amphibian ancestry.

The reptiles expanded during the Permian, the last period of the Paleozoic Era, but it was during the following Mesozoic Era, from about 225 million to 65 million years ago, that they came to dominate not only the land but also the sea and air. During this era, known as "the age of reptiles," they radiated out into many diverse forms and occupied a wide array of ecological niches. The dinosaur group, if still regarded as reptiles, included the largest land animals that ever lived. The age of reptiles came to a close at the end of the Cretaceous Period, the last period of the Mesozoic Era. The undetermined reasons for this decline have been the subject of intensive study for many years

Birds in Relation to Reptiles

Some reptiles are more closely related to birds than they are to other reptiles. Crocodiles are more closely related to birds than they are to lizards. Theropod dinosaurs are even more closely related, because birds evolved from them.

Cladistic writers prefer a more unified (monophyletic) grouping. This puts the birds (over 10,000 species) with what people normally call reptiles.

Taxonomy

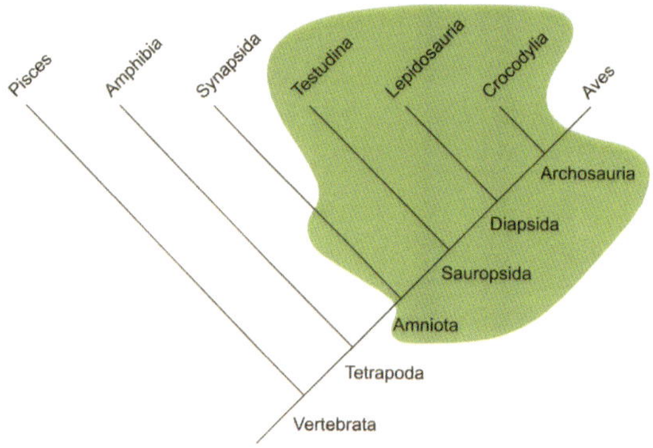

Diagram of traditional Reptilia: it is *not* a clade. To be a clade it would need to include birds (Aves) and drop Amniota. This is why the term 'Sauropsida' is often preferred to the term 'Reptilia' (in modern taxonomy).

Reptilia is an evolutionary grade rather than a clade. The main reason is that the term 'reptile' does not include birds, the descendents of theropod dinosaurs. Another reason is that the word 'reptile' is misleading because many extinct types were very different from living reptiles.

So instead of Reptilia as a taxonomic class, today most experts use Class Sauropsida (which includes all reptiles and birds, living and extinct). Class Synapsida includes mammals and all their forebears. *Reptile* is still the usual informal term to describe living snakes and lizards. Mammals are a genuine clade, and so Mammalia is still the taxonomic term.

Since reptiles are not monophyletic, reclassifying them is one of the key aims of researchers. Some taxonomists, such as Benton, make Sauropsida and Synapsida as class-level taxa. The two groups split in the Carboniferous, from stem-group Amniotes (the early tetrapods, which laid cleidoic eggs).

Living Reptiles

- *Reptiles* (living groups only)
    - Order Crocodilia (crocodiles, gavials, caimans, and alligators): 23 species
    - Order Sphenodontia (*Tuatara* from New Zealand): 2 species
    - Order Squamata (lizards, snakes, and amphisbaenids ("worm-lizards"): about 7,000 species.
    - Order Testudines (turtles, tortoises, and terrapins): approximately 300 species

Amphibian

The Class Amphibia (or amphibians) includes three Orders:

- Anura (frogs and toads) – about 6,500 species
- Caudata or Urodela (newts and salamanders) – about 680 species
- Gymnophiona or Apoda (caecilians) – about 205 species

Frogs and Toads

Newts and Salamanders

Caecilians

Most people are familiar with frogs, toads, newts and salamanders, but are less familiar with caecilians, which superficially resemble large earthworms. Little is known about most caecilians as they mostly live hidden underground or underwater.

All amphibians are cold-blooded animals and most metamorphose from a juvenile to an adult form. They cannot generate their own body heat, instead relying on the temperature of their environment to help them keep warm or cool enough to survive. There are over 7,400 species of amphibians, and they inhabit all continents except Antarctica, living in varied habitats such as rainforests, rivers and streams, deserts and alpine environments.

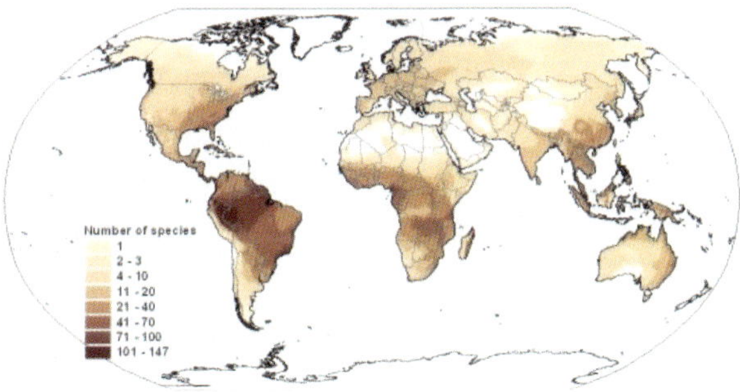

Global diversity of amphibians (from Global Amphibian Assessment).

Amphibians are important and beneficial in many ways:

- They play an important role in nature as both predator and prey, sustaining the delicate balance of nature.
- They eat pest insects, benefiting successful agriculture around the world and minimizing the spread of disease, including malaria.
- The skin of amphibians has substances that protect them from some microbes and viruses, offering possible medical cures for a variety of human diseases, including AIDS.
- Frogs have had a special place in various human cultures for centuries, cherished as agents of life and good luck.

In addition, successful collaboration to save amphibians can help mankind to more confidently face and address other major environmental challenges.

One-third to one-half of the world's approximately 7,500 known amphibian species could go extinct in our lifetime. This would be the largest mass extinction since the disappearance of the dinosaurs. Amphibians are considered canaries in the coal mine: they are among the first species to be affected by environmental stressors, so when they show declines in the wild, it is a warning to other species, including humans. Their population declines and extinctions signal that changes are occurring in the environment that will also negatively impact humans.

Amphibians are dying in alarming numbers. About 122 species are believed to have become extinct since 1980. About half of the world's 7,500 species could become extinct within our lifetime with at least 500 species facing imminent extinction. This rate of extinction is unprecedented since the demise of the dinosaurs.

Destruction of habitat, invasive species, pollution, global warming and other human influences threaten many species throughout the animal and plant kingdom. In addition to these challenges, amphibians are being attacked by a lethal disease that has caused unprecedented and unexpected losses, called chytrid fungus. Chytrid quickly and quietly destroys entire species when introduced to new populations in the wild. Chytrid fungus has now been identified in association with amphibian die-offs on every amphibian-inhabited continent. This disease has decimated or wiped out species that had not been considered critically endangered. It happens so quickly that species disappear before anyone realizes they are in trouble.

Amphibian chytrid is a disease that infects the skin of amphibians, a vital organ through which many drink and breathe. It was discovered a decade ago; dozens of frog species have already vanished because of it. In environments where it thrives, the fungus can kill 80 percent of the native amphibians within months. Currently, it is unstoppable and untreatable in the wild, even in 'protected' areas. Amphibian chytrid is believed to have originated in Africa. The export of African clawed frogs (likely resistant carriers of the fungus) around the world for human pregnancy testing and lab studies spread this disease worldwide. Recently, the food and pet trades may have contributed to the problem as well. The chytrid's spread and effects may be exacerbated by climate change – warmer temperatures dry the moist areas where amphibians live, causing stress that may lead to greater susceptibility to the disease.

For amphibians, the potential impact of chytrid fungus across all species is probably underestimated, and a significant number of amphibian species that have been described are too rare and too poorly known to assess their level of endangerment.

While there are 2,000 to 3,000 threatened amphibian species, many suffer from threats that we have the ability to mitigate in the wild, such as habitat destruction and over-collection. AArk is using a species conservation needs assessment tool and regional expertise in a series of workshops around the world to determine which species require our intervention.

Anura

The order Anura includes the frogs and toads. There is no fundamental difference between frogs and toads. Frogs have a short body, webbed digits (fingers or toes), protruding eyes, forked tongue and no tail. They are exceptional jumpers: many of their features, particularly their long, powerful legs, are adaptations to improve jumping performance. They often live in semi-aquatic or inhabit humid areas.

Frogs such as this edible frog has smooth (soft) wet skin

This Texas toad has rough (hard) dry skin

A popular distinction is often made between frogs and toads on the basis of their appearance. Toads' warty skin is an adaptation for making their toxic slime. Apart from these glands, their skin is dry, and that is an adaptation to drier habitats. These features have evolved a number of times independently: convergent evolution. The distinction has no taxonomic basis. The only family exclusively given the common name "toad" is Bufonidae (the "true toads"), but many species from other families are commonly called "toads".

Caudata

This fire salamander has yellow and black stripes along its body: typical warning colouration.

Chinese fire-bellied newts has red stripes on their front body, which they can raise when attacked. This is also warning colouration.

The order Caudata is the salamanders.

Newts are salamanders which spend their life in the water even though they are adults. They are classified in the subfamily Pleurodelinae of the family Salamandridae.

Respiration differs between species of salamanders. Species that lack lungs respire through gills. In most cases, these are external gills, visible as tufts on either side of the head. Some salamanders that are terrestrial have lungs that are used in respiration, although these are simple and sac-like, unlike the more complex organs found in mammals. Many species, such as the Olm, have both lungs and gills as adults.

Some terrestrial species lack both lungs and gills and perform gas exchange through their skin. Even some species with lungs also respire through the skin in this manner.

The skin of salamanders secretes mucus. This helps to keep the animal moist when on dry land, keeps their salt balance while in water, and lubricates during swimming. Salamanders also secrete poison from glands in their skin, and some additionally have skin glands for secreting courtship pheromones.

Axolotls, from the genus *Ambystoma* (or mole salamanders), are neotenic amphibians. This means they get to sexual maturity and reproduce while still in a larval form.

Defence Mechanisms

Most salamanders and newts have some defence against predators, usually a poison which makes them uneatable. Their bright colours are warning colouration. If, instead, they are camouflaged, this means they are probably not protected by a toxin.

The second line of defence is to shed their tail, which can grow again. The tail wriggles a bit, attracts the predator while the business part of the salamander moves off.

Other Characteristics

There are over 350 lungless salamanders. Most of them are terrestrial and are active in daytime. Lungless salamanders may communicate with their nose. Slender salamanders are found in the Pacific Coast. They are sometimes called "worm salamanders". This is because they have slimmer (skinny) bodies than most salamanders. If touched, slender salamanders will bounce on the ground and then run away.

Gymnophiona

Beddome's Caecilian, like all other caecilians, do not have feet or limbs.

The order Gymnophiona includes the caecilians. These are long, cylindrical, limbless animals that look like snakes or worms. Their skin has circular folds, increasing their similarity to the segments of earthworms. Some are aquatic but most live underground in burrows they hollow out. Many caecilians give birth to live young, and in the animals that do not do this, the eggs may undergo metamorphosis before they hatch. Caecilians are found in tropical Africa, Asia and Central and South America. There are 171 different species.

They are burrowing amphibians. This means that they dig themselves in wet soil like worms. Their heads are strong and have bones that help them dig. Because caecilians have a lot of vertebrae, they can bend easily.

Reproduction

Amphibians are the only vertebrates to go through metamorphosis. This means that their young look different from their adult. Amphibians usually reproduce in early spring to late summer, though some reproduce in winter and fall. Most frogs and toads, such as the common frog (*Rana temporalis*), gather in large groups to ponds, rivers, swamps and lakes to breed. Male frogs and toads may croak to attract a female. When a female frog has chosen a mate, the male frog hops on top of her. They swim together as she lay eggs in the water. Sometimes, males fight to mate with a female. Frogs can lay up to 100 to 60,000 eggs in one clutch. This is called "frogspawn".

A pair of *Hylarana aurantiaca* mating

It is a fundamental feature of amphibia that their reproduction is, one way or another, tied to water. This is because their eggs, although covered by jelly, cannot survive long in dry conditions.

Eggs

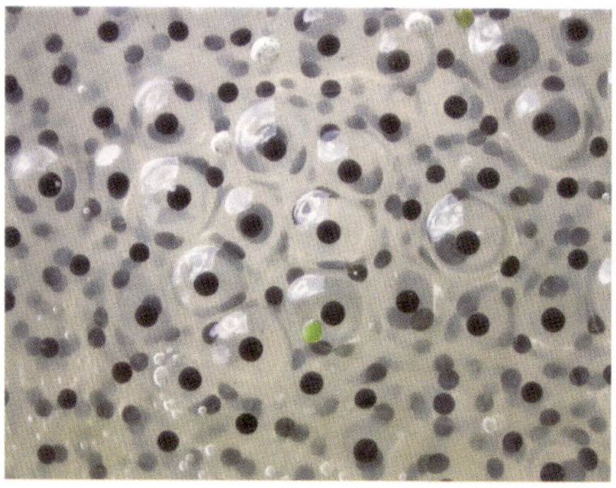

Frogs eggs are called "frogspawn"

Some amphibians lay eggs that are very clear. This makes it easy to watch a tadpole grow inside its egg

Most female amphibians lays her eggs in water. Males release sperm to fertilize them. The eggs are laid one by one or in batches. Batches of eggs can look like a long chain or a ball of foam. They may wrap their eggs around plants in the water. They do this so their eggs will not drift away.

Tree frogs usually lay their eggs on a leaf in a rainwater pool. Bullfrogs, such as the male American bullfrog and the male African bullfrog, stay with their tadpoles and protect them from predators.

They also move their tadpoles by using their nose to dig a channel to another place where there is more water. They do this so their tadpoles do not dry up. Most amphibians leave their eggs to look after themselves. Fish and other animals eat most of their eggs. Male midwife toads carry their eggs on their backs. When they are ready to hatch, the toad goes back to the water and release them.

Tadpoles

This orange-thighed Frog tadpole has only a head and a tail. It will grow back legs and front hands in a couple of weeks. It will then loose its tail and become a young frog.

Tadpoles do not have lungs when they hatch and instead have gills. Because gills have a large surface area, tadpoles can get more oxygen by using them. Young tadpoles have their gills exposed. When they get older, their gills are covered over by skin. When they hatch, tadpoles eat constantly. The tadpoles eat what is left of their eggs, this is usually their first food.

Frog, toad and newt tadpoles eat plants such as algae and pondweed or filter feed. When they get older, they may start to feed on tiny animals in the water. Salamander tadpoles and surinam horned toad tadpoles are carnivorous throughout their tadpole stage. Surinam horned toad tadpoles are very aggressive. They eat other tadpoles if food is nowhere to be found. The eggs of the spadefoot toad hatches in three days. Their tadpoles complete their metamorphosis in six to eight days. This is because spadefoot toads lay their eggs in places where water will dry up soon.

Tadpoles of frogs and toads start to grow their back legs first. They then grow front legs a few weeks later. When tadpoles grow their limbs they are called "froglets". This is because they look rather like a smaller version of adult frogs and toads. Tadpoles will also start to grow a backbone after growing their front limbs. After this, their mouths get bigger and their eyes will stick out more. After a tadpole has grown its hands, their tails continue to get shorter until there is nothing left of them.

Habitats

Salamanders and newts can be found living in streams. Salamanders can be found in rotten logs, holes or underground places that are wet such as under leaves. Web-toed salamanders live in habitats where there are a lot of rocks. They like to hide under rocks and stones. The tailed frogs, like to live in cold water habitats. In their habitat, amphibians like to live where there are a lot of places to hide. These include nearby small trees, logs and plants. While underwater they like to hide near aquatic plants and rocks. Tree and dart frogs like to live in forests on trees, plants and on the ground under leaves.

Most frogs like to spend their time in the water hiding near aquatic plants

This Alpine newt is basking on a rock. It does this to get heat from the sun

Some amphibia can be found living in the desert or the arctic. The desert froglet lives in the desert. They are only active at night, when temperatures are much cooler. It rarely rains in the desert and because of this, desert frogs will burrow to keep cool. They use their mucus to keep them wet. They will spread it all over their bodies. The mucus will harden to keep the water it produces from escaping. Once the desert frog has done this, it will stay in its cocoon and will not move. They will stay like this for several months to years until a rainstorm. Desert frogs and toads lose water more quickly. The spadefoot toad will spit on the ground. Once they have done this, they will lay on it. Their bodies will take in the water. Their bodies are thin and have a lot of blood vessels, this helps them to be able to take water through their skin. The California newt can survive a fire by spreading its mucous over its body.

Arctic frogs such as the wood frog, moor frog and the common frog has to live with freezing temperatures for a long time. They will burrow in places where they can get into a cocoon. Like every living organisms, amphibians must have water to survive. Amphibians however, need freshwater. Some frogs such as burrowing frogs can keep water in their bladders. This allows them to stay underground without drying up. The crab-eating frog lives near water that is somewhat salty. They will eat saltwater crabs. Torrent salamanders lives in cold waters. Because of this they have shorter lungs. Short lungs helps them to float easily.

Distribution

Amphibia are world-wide, though restricted in distribution by their need for moist or watery habitats to reproduce.

Anatomy

Skin

This poison dart frog's skin is blue. This color warns animals that they are toxic

Most amphibians have a good sense of smell. Their eyes have color and clear vision

Many amphibia have secretions in their skin which makes them toxic. They do not produce toxins themselves. They get toxins from what they eat. They eat insects in their habitat. These insects get the poison from a plant. The toxin has been discovered in beetles. This means that they are likely the cause of poisons found in amphibians. Amphibians do not produce batrachotoxin in captivity, which means that it is not harmful to touch them. The American Indian tribe comechingóns used the toxins of the arrow dart frogs when hunting.

Newts in the genus *Taricha* has a poison called tetrodotoxin, a neurotoxin. Scientists believe that toxins in newts are caused by bacteria in the genera *Pseudoalteromonas*, *Pseudomonas* and *Vibrio*. Because of this, the newts do not have a lot of predators. However, some species of snakes have develop a resistance. This means that they can eat newts without the toxin hurting them. It is a case of co-evolution.

Senses and Skeletal System

Amphibians' eyes have lids, glands and ducts. They have good colour vision Caecilian eyes are small and dark. Most of them are blind. Most amphibians have a good sense of smell, even underwater.

The skeletal system of amphibians are similar to other four-legged animals. They have a spine, rib cage, long bones such as the humerus and femur. They also have short bones such as the phalanges, metacarpals, and metatarsals. Most amphibians have four limbs, except for caecilians. The bones in amphibians are hollow and do not weigh much.

Diet

Amphibians are predatory animals. If there is no food to be found, they will eat each other.

Amphibians are predatory animals. They mostly eat live invertebrates and animals that do not move too quickly. These include caterpillars, earthworms, crayfish, water beetles, snails and dragon fly larvae. Many amphibians use their sticky tongues to catch their prey. They will swallow the animal whole, but may chew it just a bit for it to go down their throats. The Ranidae family and the Ceratophrys genus will eat almost anything they can fit into their mouths. These include rodents, birds, ducklings, small fish and small mammals. Most frogs are cannibalistic, and will eat each other if food is no where to be found. Some amphibians will even eat their own tadpoles and eggs if there is no food for them.

Feeding in Captivity

In captivity, pet frogs will be given crickets, worms, small fish, rodents and fruit flies. Adult amphibians can help decrease the mosquito population by eating most of their larvae.

Caecilians eat earthworms, termites and beetle larvae, and also small lizards. Caecilians rely on their smell to find food. They like to eat earthworms and will find them by picking up their chemical signals. Salamanders and newts are fed a lot of different types of worms. These include blood worms and earthworms. They can eat small fish such as goldfish, fathead minnows and guppies. Salamanders also eat crickets and pinkies, which are baby rats.

Conservation

The golden toad of Monteverde

The amphibian population have been decreasing from all locations in the world. Scientists have said that the declining of amphibians is one of the most critical threats to global biodiversity. A number of causes are believed to be involved. These include habitat destruction, over-exploitation, pollution, introduced species, climate change, destruction of the ozone layer, and diseases like chytridiomycosis. Ultraviolet radiation damages the skin, eyes and eggs of amphibians. However, the declines of amphibian population are still not understood.

The Amphibian Conservation Action Plan (ACAP) have released a global strategy to help the amphibian population. It was developed by over 80 leading experts. The Amphibian Specialist Group of the World Conservation Union (IUCN) are working on another global strategy to help amphibian populations. The Amphibian Ark (AA) is an organization that was created to help the public be aware of the decline in amphibian populations. They have been working with zoos and aquaria around the world. They try to encourage them to create a natural habitat for threatened amphibians. Another project is the Panama Amphibian Rescue and Conservation Project (PARCP) which are trying to spread awareness about chyridiomycosis. The disease is spreading into eastern Panama and threatening all amphibians living there.

On January 21, 2008, Evolutionarily Distinct and Globally Endangered (EDGE) released a statement to the public. It was created by Helen Meredith, who identified nature's most endangered species. Meredith explains that 85% of the top 100 endangered amphibians list are receiving little to no conservation attention.

Human use

As Food

Bullfrogs being sold alive at a supermarket in China.

Bullfrog legs are a source of food for Southern United States and the Midwestern United States. People hunt bullfrogs at night near rivers. The bullfrogs' legs are cooked, while their backs are fried. In China, bullfrogs are sold alive for eating. However, they are later cooked dead with vegetables. In the state of California, people must have a license to catch bullfrogs for food. In schools, bullfrogs are dissected in biology classes. Usually, this is done in grammar school. The dissecting is a method for teaching students the anatomy of a bullfrog. The emperor newt is hunted in China for food. They are also used there for medicine. Burrowing frogs are able to hold water in their bladder, because of this indigenous Australians use them to drink water.

As Pets

Amphibians are also kept as pets. They are kept in aquariums or a terrarium. A terrarium is a tank that is decorated with plants and soil on one side. On the other side, there is water. Most amphibians would need one place for land and another for water. Each type of amphibian should have its special needs taken care of. Semi-aquatic amphibians need both land and water divided in the tank. Tropical frogs would need mist and high humidity in their terrariums. Water for amphibia needs dechlorination. The chlorine in tap water can kill amphibia. Some amphibians popular exotic pets, and are found in pet stores that sell reptiles.

References

- Gudger, E. W. (1924). "Pliny's Historia naturalis. The Most Popular Natural History Ever Published". Isis. 6 (3): 269–281. doi:10.1086/358236. ISSN 0021-1753
- Hebert PD, Stoeckle MY, Zemlak TS, Francis CM (2004). "Identification of Birds through DNA Barcodes". PLoS Biology. 2 (10): e312. doi:10.1371/journal.pbio.0020312. PMC 518999 . PMID 15455034
- Gill, Frank (2006). Birds of the world: recommended English names. Princeton: Princeton University Press. ISBN 978-0-691-12827-6
- "Techniques to capture Seaducks in the Chesapeake Bay and Restigouche River". USGS. Retrieved 2007-12-01
- Beddall, Barbara G. (1957). "Historical notes on avian classification". Systematic Zoology. 6 (3): 129–136. doi:10.2307/2411751
- White, Jeanne A. (1999). "Ornithology Collections in the Libraries at Cornell University: A Descriptive Guide". Retrieved 2007-12-01
- Clack Jennifer A. 2012. Gaining Ground: the origin and early evolution of tetrapods. 2nd ed, Indiana University Press, Bloomington IN. ISBN 978-0-253-35675-8
- Greenwood, J.J.D. (2007). "Citizens, science and bird conservation" (PDF). Journal of Ornithology. 148 (1): 77–124. doi:10.1007/s10336-007-0239-9. Archived from the original (PDF) on 2011-07-19
- Brands, Sheila (14 August 2008). "Systema Naturae 2000 / Classification, Class Aves". Project: The Taxonomicon. Retrieved 11 June 2012
- Stresemann, Erwin (1975). Ornithology. From Aristotle to the Present. Cambridge, Massachusetts: Harvard University Press. pp. 170–191. ISBN 0-674-64485-9
- Birkhead TR & S van Balen (2008). "Bird-keeping and the development of ornithological science". Archives of Natural History. 35 (2): 281–305. doi:10.3366/E0260954108000399
- van Oppenraay, Aafke M.I. (2017). "Avicenna's Liber de animalibus(Abbreviatio Avicennae) - Preliminaries and State of Affairs" (PDF). Documenti e studi sulla tradizione filosofica medievale. 28: 401–416. Retrieved 4 May 2018
- BirdLife International (2000). "Threatened Birds of the World: The official source for birds on the IUCN Red List". Lynx Edicions, Barcelona, and BirdLife International, Cambridge, UK. ISBN 0946888396
- Koinm, Albert J. (2000). "Christopher Merrett's Use of Experiment". Notes and Records of the Royal Society of London. 54 (1): 23–32. doi:10.1098/rsnr.2000.0093
- Johnson, Kristin (2004). "The Ibis: Transformations in a Twentieth Century British Natural History Journal". Journal of the History of Biology. 37 (3): 515–555. doi:10.1007/s10739-004-1499-3
- Campbell, Jonathan A. 1999. Amphibians and reptiles of Northern Guatemala, the Yucatán, and Belize. University of Oklahoma Press, ISBN 0806130660
- McCallum, M.L. (2007). "Amphibian decline or extinction? Current declines dwarf background extinction rate". Journal of Herpetology 41 (3): 483–491. doi:10.1670/0022-1511(2007)41[483:ADOECD]2.0.CO;2

Chapter 3

Invertebrate Zoology

Invertebrate zoology is a subdiscipline of zoology, which studies animals that do not have a backbone. Arthropodology, Cnidariology, Helminthology and Malacology are some of the subdivisions of invertebrate zoology. They have been elaborated in this chapter to develop a better perspective about the branches of invertebrate zoology.

Invertebrate zoology (IZ) is the study of animal biodiversity. "Invertebrates" include all animals exclusive of vertebrates, in an artificial division set up centuries ago. Of the 34 or so major divisions (phyla) of animal life, 33 and 2/3 are "invertebrate". Animals come in a bewildering diversity of forms and pursue an amazing breadth of life styles, including:

Sponges live attached to the ocean floor, lack a nervous system, and filter water for food. Except some deep sea species turned to snagging plankton with specialized, hooked spicules.

Corals build reefs visible from space, the largest structures created by living things, including humans. Internal microscopic algae feed them, but they also indulge in carnivorous fare. Corals host countless other species in their complex colonies.

Worms... The word may repulse some, but for a zoologist, it means diversity. Over half of the animal phyla are "worms" - elongate, muscular organisms that swim in the ocean, burrow in the sea floor or reef, live in soil, or crawl inside most plants and animals large enough to host "helminth" parasites. Feather duster worms, pin worms, large flat worms that swim like flying carpets are unrelated "invertebrates" all sporting a "wormy" body plan.

Mollusks, the second largest phylum, are big not only in species diversity but in body form: worms, clams, slugs, flying squid, tusk shells, chitons, limpets, sea butterflies. FLMNH hosts about a third of the 100,000 known species.

More species of arthropods are known than of all other animals put together. A better division of the world would be into Arthropods and Unarthropods... Body forms range from parasitic, pulsating blobs, to centipedes, horseshoe crabs, and butterflies.

Echinoderms include the conspicuous sea stars, sea cucumbers, and sea urchins. They are our symmetrical, large, striking, but weird cousins. How many of your other cousins can change the consistency of their body to turn to slime, or to lock their arm in a position so that it will break before unbending?

Chordates are the most familiar animals, as they include fish, frogs, eagles, and humans. But they also include lancelets, salps, larvaceans, and sea squirts. Animals that make mucus houses several times their own size for filter feeding. Bioluminescent colonies 30 feet long that jet propel through the ocean. Brainless, fish-like lancelets that burrow in the ocean floor to filter-feed on plankton.

Invertebrate

Invertebrates are animals without a backbone. Of the planet's estimated 15-30 million animal species, 90% or more are invertebrates. Invertebrates live just about anywhere. They have been recorded in the upper reaches of the atmosphere, in the driest of deserts and in the canopies of the wettest rainforests. They can even be found in the frozen Antarctic or on the deepest ocean floor.

Colourful Lynx Spider, *Oxyopes*

There is a good chance that you have seen an invertebrate recently. Do you recall batting away a fly, unearthing a worm, or admiring a spider as it waited to catch food in its web. Well guess what? All of these animals and many more are collectively known as invertebrates - animals that lack a backbone.

Invertebrates are Everywhere

There are so many invertebrates on this planet that it is impossible to count them all. They come in many shapes and sizes, live practically anywhere and provide many services that are vital for our survival.

Invertebrates have been recorded in the upper reaches of the atmosphere, in the driest of deserts and in the canopies of the wettest rainforests. They can even be found in the frozen Antarctic (some mites and springtails can withstand temperatures of -35°C) or on the ocean floor of the Abyssal Sea which reaches depths of up to 11 000 metres.

Invertebrates are all around us and yet amazingly most go about their daily business unnoticed. Much of this has to do with the size of invertebrates. On land, invertebrates range from fractions of a millimetre to approximately 150 centimetres in length, though most are less than five centimetres. Their size allows them to occupy large habitats such as deserts and rainforests as well as many microhabitats, which may not be visible to the eye. For example, they maybe found in a pool of water in a pitcher plant, the follicles of a hair, or inside a seed.

Invertebrate Groups

Terrestrial (land) invertebrates include the following groups, many of which also have members that live in freshwater or marine environments:

- Insects
- Spiders
- Centipedes
- Millipedes
- Worms
- Velvet worms
- Slaters
- Landhoppers

Marine and freshwater invertebrates include the following groups, some of which have land-dwelling members also:

- Snails and slugs
- Sea stars and sea urchins
- Anemones and corals
- Crabs, lobsters, prawns and crayfish
- Sponges
- Bluebottles, jellies and comb jellies

Divisions of Invertebrate Zoology

Invertebrates represent 97% of all named animal species, and because of that fact, this subdivision of zoology has many further subdivisions, including but not limited to:

- Arthropodology - the study of arthropods, which includes
    - Arachnology - the study of spiders and other arachnids
    - Entomology - the study of insects
    - Carcinology - the study of crustaceans
    - Myriapodology - the study of centipedes, millipedes, and other myriapods

- Cnidariology - the study of Cnidaria
- Helminthology - the study of parasitic worms.
- Malacology - the study of mollusks, which includes
  - Conchology - the study of Mollusk shells.
  - Limacology - the study of slugs.
  - Teuthology - the study of cephalopods.
- Invertebrate paleontology - the study of fossil invertebrates

These divisions are sometimes further divided into more specific specialties. For example, within arachnology, acarology is the study of mites and ticks; within entomology, lepidoptery is the study of butterflies and moths, myrmecology is the study of ants and so on.

Arthropodology

Arthropodology is a biological discipline that focuses on the arthropods. Arthropods are invertebrates that belong to the phylum Arthropoda. This phylum is the largest among the phyla in the animal kingdom. They include the insects, the arachnids, the myriapods and the crustaceans. These animals are grouped under this phylum based on the features common to them. They have multiple paired jointed limbs and a hard chitinous exoskeleton. The exoskeleton is a hard outer casing that protects their soft internal organs. Arthropodology deals with these animals and aims to understand their evolution, anatomy, biology, and ecological importance.

Arthropodology that studies the parasitic mode of life of certain arthropods is referred to as medical arthropodology. Apart from parasitic arthropods, there are also arthropod species that act as vectors of disease. An example of that would be the mosquito. Certain species of mosquitoes serve as vectors for disease-causing viruses and parasites. Mosquito-borne viral diseases include the yellow fever, dengue fever, and zika. Mosquito-borne parasitic diseases are malaria and lymphatic filariasis.

Arthropodology may be subdivided into the following fields based on the animal groups that they study: arachnology (studies spiders and other arachnids), entomology (studies insects), carcinology (studies crustaceans), and myriapodology (studies myriapods).

Arthropod

Arthropods Definition

An "arthropod" is an invertebrate animal that has an exoskeleton, a segmented body, and jointed appendages. The following families of organisms are all examples of arthropods:

- Insects such as ants, dragonflies, and bees
- Arachnids such as spiders and scorpions

- Myriapods (a term which means "many feet") such as centipedes and milipedes
- Crustaceans such as crabs, lobsters, and shrimp

It may help to remember that the term "arthropod" comes from the Greek words for "jointed foot." If the organism has an exoskeleton with joints between its feet and its body, it is probably an arthropod!

Arthropods are a lineage of life that developed skeletons on the outside – their hard shells, made of a material called "chitin" – instead of on the inside for structural support.

Arthropods' bodies also have other important differences from those of vertebrates like ourselves – their organ systems are simpler and less efficient, which limits the size arthropods can attain.

An ant the size of a human, for example, would not be able to pump oxygen through its blood to feed all its tissues, since the arthropod circulatory system is simpler and less efficient than a humans'.

All arthropods are thought to have evolved from a single common ancestor, though scientists are not sure what this common ancestor looked like, or exactly when it lived.

Arthropod Characteristics

Characteristics shared by all arthropods include:

- Exoskeletons made of chitin
- Highly developed sense organs
- Jointed limbs (the limbs must be jointed like the joints in a suit of armor, since the exoskeleton is rigid and cannot bend to allow movement)
- Segmented bodies
- Ventral nervous system. "Ventral" means "in front," so this means that arthropods' nervous systems run along the front of their bodies, near their stomachs, instead of along their backs like the spinal cords of animals.
- Bilateral symmetry. This means that the left and right sides of an arthropod are the same – it will have the same number and arrangement of legs, eyes, etc. on the right side of its body as on the left.

Types of Arthropods

Trilobites

Trilobites were an ancient family of marine arthropods that went extinct during the Permian-Triassic extinction event. Today, they are known to us mostly through fossils like the one below.

They lived on the ocean floor and occupied ecological niches similar to those occupied by crustaceans today.

Asaphus platyurus

Chelicerates

Chelicerata are a branch of the arthropod family tree that, at first glance, may not appear related to each other.

This family includes arachnids (such as spiders and scorpions), sea spiders (which look similar to arachnids but have some important differences), and horseshoe crabs (which, despite their name, have important differences from other crustaceans).

Myriapods

The term "myriapod" means "many legs" – so it is not surprising that centipedes, milipedes, and other many-legged creatures are part of this family.

Myriapods can have anywhere from less than ten legs – to over 750! That just seems excessive.

Myriapods are typically found in forests and other ecosystems where there is lots of decaying plant and animal material for them to feed on.

Crustaceans

Crustaceans are a family of primarily aquatic arthropods that include lobsters, crabs, shrimp, crayfish, barnacles, and the odd one out – wood lice, also known as pill bugs or "roly polys."

Unlike their aquatic cousins, wood lice live mostly on dry land and are found in environments such as gardens and forests, where they survive by eating decaying plant and animal material.

It may also surprise you to see barnacles included on this list: adult barnacles develop hard shells that stick them to their surroundings, such as the bottoms of boats or other underwater surfaces.

But earlier in their lives before they freeze in place, barnacles have bodies with legs much like the other crustaceans!

Hexapods

The term "hexapod" literally means "six feet." It might not surprise you to learn that insects – which all have six legs – are hexapods.

Insects include most "bugs" with six legs, such as flies, ants, termites, beetles, dragonflies, mosquitoes, cockroaches, butterflies, and moths.

There are also three much smaller groups of animals in the "hexapod" category. Collembola, Protura, and Diplura were all once considered to be insects, but later found to have small differences that set them apart from other insects.

Examples of Arthropods

Ants

When you think of a stereotypical arthropod body, you probably think of an ant. Ants have hard exoskeletons and jointed legs. They also have bodies which are clearly segmented into a head, thorax, and abdomen.

Ants show one type of social organization that has been developed by arthropods. Ants, bees, and termites are all what is called "eusocial" organisms – organisms living in extreme degree of cooperation, with "colonies" that almost operate like a single organism themselves.

Most arthropod species are not eusocial, but eusocial colony life is one of the fascinating roads that arthropod evolution has taken.

Spiders

Spiders are also arthropods, possessed of hard exoskeletons, segmented bodies, and jointed limbs.

Spiders typically eat smaller arthropods, such as gnats and flies – though they will eat any living thing they can catch, and some particularly huge spiders have been known to eat birds or rodents!

Spiders have evolved a variety of strategies for catching their prey – some spin sticky, nearly invisible webs that prey animals wander into and get stuck. Others are active hunters, including jumping spiders which can jump at extreme speeds using special mechanisms in their legs.

Some spiders combine these two strategies, such as "trap door" spiders, which set traps by creating hiding places for themselves – and then jumping out to grab unsuspecting prey animals that wander by!

Lobsters

With lobster being considered a luxury food today, it's easy to forget that lobsters are in the same family as spiders and ants.

Crustaceans can grow bigger underwater than on land – and lobsters can grow to weigh nearly 50 pounds!

Lobsters' body design has changed little in the last 100 million years, and their anatomy is spectacularly weird. The lobster's kidneys are located in its head, its brain in its throat, and its teeth in its stomach. Its "ears" for picking up sound are located in its legs, and its tastebuds, like those of insects, are in its feet.

Butterflies

Butterflies are the most famous example of arthropod metamorphosis.

At some point in their lifecycle, all arthropods go through a drastic change from their larval stage to their adult form. But butterflies are the only ones whose adult forms are so beautiful that we pay attention to this change.

The common features of exoskeleton, jointed limbs, and segmented body can be seen in adult butterflies.

Facts About Arthropods

- Arthropods colonized land about 100 million years before vertebrates did. It's thought that colonizing land was easier for them for several reasons – including the fact that they had already evolved legs, which they used for walking on the bottom of the sea.
- About 80% of all animal species are arthropods! We don't see them very often in our daily lives, but all the species of bugs and crustaceans on Earth add up!
- All arthropods undergo metamorphosis – a process where their bodies change radically as they pass from their larval to adult stages. Butterflies are the best-known for entering cocoons as caterpillars and coming out quite different, but all arthropods do something similar!
- When arthropods outgrow their old exoskeleton, they have to molt – leaving behind their former skin and growing a new one. All arthropods have to do this at least once in their lives.
- Crustaceans and arachnids – two types of arthropods – have blue blood instead of red blood!This is because their blood uses a blue copper compound to carry oxygen, instead of the red iron compound used by animals.
- Arthropods' hard exoskeletons is made of chitin – which is made of a derivative of the sugar glucose! But chitin would not taste sweet, and you wouldn't be able to eat it; to make it hard and strong, the glucose is modified so that our bodies no longer recognize it as sugar.

Subdivisions of Arthopodology

Arachnology

Arachnology is the scientific study of spiders and related animals such as scorpions, pseudoscorpions, and harvestmen, collectively called arachnids. Those who study spiders and other arachnids are arachnologists.

Nephila clavipes

Arachnology as a Science

Arachnologists are primarily responsible for classifying arachnids and studying aspects of their biology. In the popular imagination they are sometimes referred to as 'spider experts'. Disciplines within arachnology include naming species and determining their evolutionary relationships to one another (taxonomy and systematics), studying how they interact with other members of their species and/or their environment (behavioural ecology), or how they are distributed in different regions and habitats (faunistics). Other arachnologists perform research on the anatomy or physiology of arachnids, including the venom of spiders and scorpions. Others study the impact of spiders in agricultural ecosystems and whether they can be used as biological control agents.

Subdisciplines

Arachnology can be broken down into more specific specialties. These topics include:

- Acarology – the study of ticks and mites.
- Araneology – the study of spiders.
- Scorpiology – the study of scorpions.

Arachnid

Arachnids are a class (Arachnida) of joint-legged invertebrate animals (arthropods), in the subphylum Chelicerata. All arachnids have eight legs, although the front pair of legs in some species has converted to a sensory function, while in other species, different appendages can grow large enough to take on the appearance of extra pairs of legs. The term is derived from the Greek word (*aráchnē*), from the myth of the hubristic human weaver Arachne who was turned into a spider. Spiders are the largest order in the class, which also includes scorpions, ticks, mites, harvestmen, and solifuges.

Almost all extant arachnids are terrestrial, living mainly on land. However, some inhabit freshwater environments and, with the exception of the pelagic zone, marine environments as well. They comprise over 100,000 named species.

Morphology

Basic characteristics of arachnids include four pairs of legs (1) and a body divided into two tagmata: the cephalothorax (2) and the abdomen (3) Almost all adult arachnids have eight legs, and arachnids may be easily distinguished from insects by this fact, since insects have six legs. However, arachnids also have two further pairs of appendages that have become adapted for feeding, defense, and sensory perception. The first pair, the chelicerae, serve in feeding and defense. The next pair of appendages, the pedipalps, have been adapted for feeding, locomotion, and/or reproductive functions. In Solifugae, the palps are quite leg-like, so that these animals appear to have ten legs. The larvae of mites and Ricinulei have only six legs; a fourth pair usually appears when they moult into nymphs. However, mites are variable: as well as eight, there are adult mites with six or even four legs.

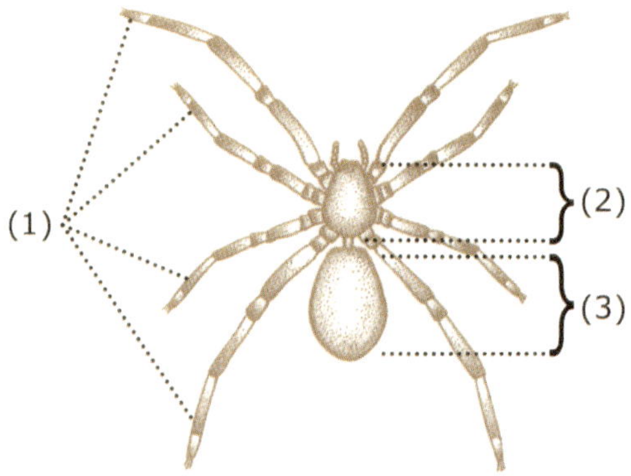

Basic characteristics of arachnids include four pairs of legs (1) and a body divided into two tagmata: the cephalothorax (2) and the abdomen (3)

Arachnids are further distinguished from insects by the fact they do not have antennae or wings. Their body is organized into two tagmata, called the prosoma, or cephalothorax, and the opisthosoma, or abdomen. The cephalothorax is derived from the fusion of the cephalon (head) and the thorax, and is usually covered by a single, unsegmented carapace. The abdomen is segmented in the more primitive forms, but varying degrees of fusion between the segments occur in many groups. It is typically divided into a preabdomen and postabdomen, although this is only clearly visible in scorpions, and in some orders, such as the Acari, the abdominal sections are completely fused. A telson is present in scorpions, where it has been modified to a stinger, and in the Schizomida, whip scorpions and Palpigradi.

Like all arthropods, arachnids have an exoskeleton, and they also have an internal structure of cartilage-like tissue, called the endosternite, to which certain muscle groups are attached. The endosternite is even calcified in some Opiliones.

Locomotion[edit]Most arachnids lack extensor muscles in the distal joints of their appendages. Spiders and whipscorpions extend their limbs hydraulically using the pressure of their hemolymph. Solifuges and some harvestmen extend their knees by the use of highly elastic thickenings in the joint cuticle. Scorpions, pseudoscorpions and some harvestmen have evolved muscles that extend two leg joints (the femur-patella and patella-tibia joints) at once. The equivalent joints of the pedipalps of scorpions though, are extended by elastic recoil.

Physiology

There are characteristics that are particularly important for the terrestrial lifestyle of arachnids, such as internal respiratory surfaces in the form of tracheae, or modification of the book gill into a book lung, an internal series of vascular lamellae used for gas exchange with the air. While the tracheae are often individual systems of tubes, similar to those in insects, ricinuleids, pseudoscorpions, and some spiders possess sieve tracheae, in which several tubes arise in a bundle from a small chamber connected to the spiracle. This type of tracheal system has almost certainly evolved from the book lungs, and indicates that the tracheae of arachnids are not homologous with those of insects.

Further adaptations to terrestrial life are appendages modified for more efficient locomotion on land, internal fertilisation, special sensory organs, and water conservation enhanced by efficient excretory structures as well as a waxy layer covering the cuticle.

The excretory glands of arachnids include up to four pairs of coxal glands along the side of the prosoma, and one or two pairs of Malpighian tubules, emptying into the gut. Many arachnids have only one or the other type of excretory gland, although several do have both. The primary nitrogenous waste product in arachnids is guanine.

Arachnid blood is variable in composition, depending on the mode of respiration. Arachnids with an efficient tracheal system do not need to transport oxygen in the blood, and may have a reduced circulatory system. In scorpions and some spiders, however, the blood contains haemocyanin, a copper-based pigment with a similar function to haemoglobin in vertebrates. The heart is located in the forward part of the abdomen, and may or may not be segmented. Some mites have no heart at all.

Diet and Digestive System

Arachnids are mostly carnivorous, feeding on the pre-digested bodies of insects and other small animals. Only in the harvestmen and among mites, such as the house dust mite, is there ingestion of solid food particles, and thus exposure to internal parasites, although it is not unusual for spiders to eat their own silk. Several groups secrete venom from specialized glands to kill prey or enemies. Several mites and ticks are parasites, some of which are carriers of disease.

Arachnids produce digestive juices in their stomachs, and use their pedipalps and chelicerae to pour them over their dead prey. The digestive juices rapidly turn the prey into a broth of nutrients, which the arachnid sucks into a pre-buccal cavity located immediately in front of the mouth. Behind the mouth is a muscular, sclerotised pharynx, which acts as a pump, sucking the food through the mouth and on into the oesophagus and stomach. In some arachnids, the oesophagus also acts as an additional pump.

The stomach is tubular in shape, with multiple diverticula extending throughout the body. The stomach and its diverticula both produce digestive enzymes and absorb nutrients from the food. It extends through most of the body, and connects to a short sclerotised intestine and anus in the hind part of the abdomen.

Senses

Arachnids have two kinds of eyes: the lateral and median ocelli. The lateral ocelli evolved from compound eyes and may have a tapetum, which enhances the ability to collect light. With the exception of scorpions, which can have up to five pairs of lateral ocelli, there are never more than three pairs present. The median ocelli develop from a transverse fold of the ectoderm. The ancestors of modern arachnids probably had both types, but modern ones often lack one type or the other. The cornea of the eye also acts as a lens, and is continuous with the cuticle of the body. Beneath this is a transparent vitreous body, and then the retina and, if present, the tapetum. In most arachnids, the retina probably does not have enough light sensitive cells to allow the eyes to form a proper image.

In addition to the eyes, almost all arachnids have two other types of sensory organs. The most important to most arachnids are the fine sensory hairs that cover the body and give the animal its sense of touch. These can be relatively simple, but many arachnids also possess more complex structures, called trichobothria.

Finally, slit sense organs are slit-like pits covered with a thin membrane. Inside the pit, a small hair touches the underside of the membrane, and detects its motion. Slit sense organs are believed to be involved in proprioception, and possibly also hearing.

Reproduction

Arachnids may have one or two gonads, which are located in the abdomen. The genital opening is usually located on the underside of the second abdominal segment. In most species, the male transfers sperm to the female in a package, or spermatophore. Complex courtship rituals have evolved in many arachnids to ensure the safe delivery of the sperm to the female.

Arachnids usually lay yolky eggs, which hatch into immatures that resemble adults. Scorpions, however, are either ovoviviparous or viviparous, depending on species, and bear live young. In most arachnids only the females provide parental care, with harvestmen being one of the few exceptions.

Entomology

Entomology is the scientific study of insects. Insects are arthropods (phylum Arthropoda) belonging to the Class Insecta. With around 925,000 described species, insects comprise the most numerous and diverse group of animals, representing more than half (about 57 percent) of all identified animal species.

Insects have many kinds of interactions with humans and other forms of life on earth, including key roles in food webs and nutrient recycling, in symbiotic relationships with flowering plants as pollinators, as producers of such human products like silk, honey, and shellac, and as agricultural pests and vectors of disease. Given these factors and the sheer numbers of insects—combined with a human nature endowed with curiosity and creativity and an often viewed role as stewards of nature—it is not surprising that entomology is an important specialty within biology.

Part of a large beetle collection.

Though technically incorrect, the definition of entomology is sometimes widened to include the study of terrestrial invertebrates in other arthropod groups or other phyla, such as arachnids, myriapods, earthworms, and slugs.

Applied Entomology

Many entomologists are employed in the study of insects that are directly beneficial or harmful to humans. Beneficial insects include those that are pollinators of agricultural crops and ornamental flowers and those imported or otherwise vital as biological control agents. The study of beneficial insects is primarily focused on their ecology and life habits, the primary concern being the understanding of how to raise them and make them more productive, or protect them from human disturbance if they are native species such as wild bees.

Conversely, much of the study of insects (and related arthropods) that directly harm human beings or their domestic animals, called medical entomology and veterinary entomology, is focused on their physiology, with the goal of developing insect controls that are effective, while minimizing undesirable side effects. For instance, many types of insecticides have been developed that target unique aspects of insect physiology and are thus considered harmless to other kinds of animals. A risk to this approach is that insecticides can also kill beneficial insects. Considerable recent effort has gone into finding biological controls that are species-specific, such as species-specific parasites and diseases, as well as genetic controls, such as the introduction of sterile insects into a population. The combination of taking into account all aspects of insect biology, available control measures, economics, and environmental considerations is known as integrated pest management.

A few insects, chiefly blood-sucking Diptera, are vectors for a wide range of deadly diseases (Meyer 2003; Bowles and Swaby 2006). Mosquitoes are especially important disease vectors, with the genus Anopheles the principle vector of malaria, Aedes aegypti the main vector of yellow fever and dengue, and other Aedes spp. carrying the causal agents of various types of encephalitis (Meyer 2003). Other well-known vectors include the tsetse fly (genus Glossina transmits protozoan agents of African sleeping sickness), black flies (spread the parasitic roundworm Onchocerca volvulus, cause of onchoceriasis), and sand flies (genus Phlebotomus are vectors of bacteria that causes Carrion's disease, and sand flies are also agent of protozoans Leishmania spp. that cause Leishmaniasis) (Meyer 2003)

Forensic entomology specializes in the study of insect ecology for use in the legal system, as knowledge of insect behavior can yield useful information about crimes. For example, the approximate time of death or whether or not a victim was alive during a fire may be determined by using facts such as at what stage of life is an insect found at the scene.

Specialization

Given the vast number and diversity of insects, many entomologists specialize in a single order or even a family of insects. A number of these subspecialties are given their own informal names, typically (but not always) derived from the scientific name of the group:

- Apiology (or melittology) - (study of) bees
- Coleopterology - beetles
- Dipterology - flies
- Heteropterology - true bugs
- Lepidopterology - moths and butterflies
- Myrmecology - ants
- Orthopterology - grasshoppers, crickets, etc.
- Trichopterology - caddis flies

With the vast numbers, diversity, and economic importance of insects, it is not surprising that entomology is rooted in nearly all human cultures from prehistoric times. However, scientific study only began as recently as the sixteenth century. Given the voluminous amount of informal and formal study of insects, only some brief overview of select key developments are presented in this section.

Early interest of humans in insects is shown in the rock painting of bees from approximately 13,000 B.C.E. Other early indicators include jewelry dated to about 1800 to 1700 B.C.E. from Crete depicting two golden bees holding a drop of honey; a painting around 1000 B.C.E. of a Scarab beetle on a wall of Rameses IX tomb; and a discussion of Ancient Egyptian beekeeping by Roman writers Virgil, Gaius Julius Hyginus, Varro, and Columella.

The first documented forensic entomology case is reported by Song Ci around 1250 C.E. in the medico-legal text book *Xiyuan Jilu*, where he describes the use of insects in the case of a stabbing near a rice field.

In 1551, zoologist Conrad Gesner published the first volume of *Historia animalium* ("History of Animals"), a work that includes some mention of insects.

In 1602, Ulisse Aldrovandi's *Animalibus insectis libri septem, cum singulorum iconibus AD vivum expressis* was published. This work was devoted to the insects and some other invertebrates. More than a half-century later, between 1662 and 1667, Jan Goedart published *Metamorphosis and historia naturalis*, illustrating, by copper plate engravings, the metamorphosis of various insects.

In 1669, microscopist Jan Swammerdam published *History of Insects*, correctly describing the reproductive organs of insects and the process of metamorphosis. The same year the anatomist Marcello Malpighi published a treatise on the structure and development of the silkworm, the first description of the anatomy of an invertebrate. Shortly thereafter, from 1696 to 1700, Antonio Vallisneri's *Dialoghi will sopra the curiosa Origine di molti Insetti* (in English, "Dialogues on the curious origin of several insects") is published; in this manuscript he, with Francesco Redi and Malpighi, contradicts the theory of spontaneous generation of maggots.

In the eighteenth century, three kinds of entomological text appeared. First, there were *illustrative works*—showy insects, often beautifully colored—whose purpose was sensual or aesthetic. An example is afforded by Maria von Merian's *Metamorphosis Insectorum Surinamenis* ("Transformations of the insects of Surinam," 1705). It actually is a masterpiece of both art and science, and Merian, "the mother of entomology," was the first to record the full life cycle of many species of butterflies and moths.

Second, there were *descriptive and systematic (classificatory) works*, usually confined to what are now known as the Insecta. In 1710, there was the first attempt at a systematic classification of insect species with the publication of John Ray's *Historia insectorum* in English. Almost half-a-century later, in 1758, Carolus Linnaeus published his groundbreaking work *Systema Naturae*, wherein the binomial nomenclature as the convention for the naming of organisms was popularized within the scientific community. In 1761, Jacob Hübner (1761–1826), the first great world lepidopterist, was born. Before Hübner, it was held that there were few genera of Lepidoptera (butterflies), a view he overthrew. His definitions of genera remain among the best of the time and so were his classifications. Also, in 1766, Moses Harris published *The Aurelian* or *Natural History of English Insects, namely Moths and Butterflies*. This was the first book on the British Lepidoptera (butterflies and moths). Harris was a pioneer in using wing venation in insect systematics. In 1771, Johann Reinhold Forster produced the first list of American insects.

A third category of entomological texts in the eighteenth century were works advancing *various subdisciplines*, such as developmental biology (life cycles), internal anatomy, physiology, and so on. These often covered other invertebrate groups. An example is René Antoine Ferchault de Réaumur's *Memoires pour Servir a L'Historie des Insectes*.

In this century, a founding work of the scientific study of entomology was Jan Swammerdam's *Biblia naturae* or "Book of Nature," which was reissued in 1737.

From the beginning of the nineteenth century, the specialist began to predominate, harbingered by Johann Wilhelm Meigen's *Nouvelle classification des mouches à deux aile* ("New classification of the Diptera") in the first year of the century. Lepidopterists were among the first to follow Meigen's lead. The specialists fell into three categories. First, there were species describers, then specialists in species recognition, and then specialists in gross taxonomy. There were, however, considerable degrees of overlap. Also, then, as now, few could entirely resist the lure of groups other than their own, and this was especially true of those in small countries where they were the sole 'expert', and many famous specialists in one order also worked on others. Hence, for instance, many works that began as butterfly faunas were completed as general regional works, often collaboratively.

In 1801, Jean Baptiste Pierre Antoine de Monet de Lamarck published *Système des Animaux sans Vertèbres ou Tableau Généraldes Classes des Ordres et des Genres de ces Animaux* ("System of invertebrate animals or general table of classes, orders and genera of these animals"). The same year, Johann Christian Fabricius commenced a series of works in which he developed an insect classificatory system based entirely on the structure of the mouthparts. In Pisa, Italy, Pietro Rossi becomes the world's first professor of entomology.

In 1806, André Marie Constant Duméril's *Analitische Zoologie*, published in both French and German, was an important text for its methodology and higher classification of insects.

Commenced in 1815 and completed in 1826, William Kirby and William Spence's *Introduction to Entomology or Elements of the Natural History of Insects* (4 vols., London: Longman) is a masterpiece that makes an outstanding contribution to entomology; it also was one of the most popular scientific works of all time.

In 1817, the pioneering work of American entomology, Thomas Say's *American Entomology*, is published in Philadelphia. In 1859, the Entomological Society of Philadelphia was established, later (1867) to be renamed the American Entomological Society. This is the earliest national organization in the biological sciences in the United States.

In 1871, Enrico Verson (1845-1927) founded the world's first silkworm experimental station in Italy.

In 1900, Walter Reed, a United States Army major, was appointed president of a board "to study infectious diseases in Cuba paying particular attention to yellow fever." He concurred with Carlos Finlay in identifying mosquitoes as the agent of transmission. In 1902, Ronald Ross was awarded the Nobel Prize for Medicine for his discovery that malaria is carried by mosquitoes. The awarding committee made special mention of the work of Giovanni Battista Grassi on the life history of the "Plasmodium" parasite (a protozoan that uses the mosquito as a host).

In 1934, Vincent B. Wigglesworth, the "Father of Insect Physiology," wrote the first book on the subject, *The Principles of Insect Physiology*.

Insect

Insects, are a class in the phylum Arthropoda. They are small terrestrial invertebrates which have a hard exoskeleton.

Insects are the largest group of animals on earth by far: about 926,400 different species have been described. They are more than half of all known living species. They may be over 90% of animal species on Earth.

New species of insects are continually being found. Estimates of the total number of species range from 2 million to 30 million.

Insects have six legs; and most have wings. Insects were the first animals capable of flight. As they develop from eggs, insects undergo metamorphosis. Insects live all over the planet: almost all are terrestrial (live on land). Few insects live in the oceans or in very cold places, such as Antarctica. The most species live in tropical areas.

Some people call all insects "bugs", but this is not correct. Only some insects are true bugs, which is a particular order of insects. People who study insects are called entomologists.

Insect Bodies

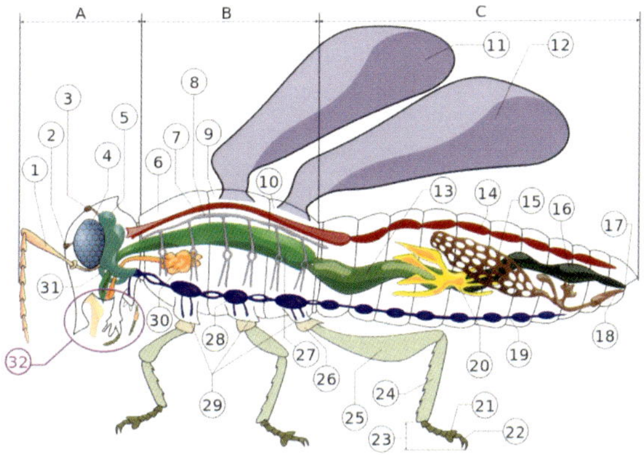

Insect anatomy

A- Head B- Thorax C- Abdomen

1. antenna 2. ocelli (lower) 3. ocelli (upper) 4. compound eye 5. brain (cerebral ganglia) 6. prothorax 7. dorsal blood vessel 8. tracheal tubes (trunk with spiracle) 9. mesothorax 10. metathorax 11. forewing 12. hindwing 13. mid-gut (stomach) 14. dorsal tube (Heart) 15. ovary 16. hind-gut (intestine, rectum & anus) 17. anus 18. oviduct 19. nerve chord (abdominal ganglia) 20. Malpighian tubes 21. tarsal pads 22. claws 23. tarsus 24. tibia 25. femur 26. trochanter 27. fore-gut (crop, gizzard) 28. thoracic ganglion 29. coxa 30. salivary gland 31. subesophageal ganglion 32. mouthparts.

Insects have exoskeletons (skeletons on the outside). Their skeletons are made out of thin, hard pieces or plates, like armour, made of chitin. All together, these pieces make a hard layer around the insect's body. The exoskeleton protects the insect.

The body of an insect has three main parts: a head, a thorax, and an abdomen. On the head are an insect's compound eyes, its two antennae (they feel and smell things), and its mouth.

On the thorax, insects have wings and legs. All insects have six legs (three pairs of jointed legs) and usually four wings (two pairs).

The abdomen is the back part of the insect. Inside the abdomen is the stomach, the heart, and the excretory system where body wastes pass out of the insect. Bees also have a stinger at the back of the abdomen.

Physiology

Just like our muscles connect to our bones to make us walk and stand up, the muscles of an insect connect to the exoskeleton to make it walk and move. Their muscles are on the *inside* of their skeleton.

Insects are cold-blooded, which means that they cannot control their body temperature. This means that insects are not good at surviving the cold, at any rate out in the open. In the winter, many insects go into something called diapause, which is the insect version of hibernation. Some insects, like cockroaches, cannot go into diapause and they will die if it gets too cold outside. This is why cockroaches love living in people's warm houses.

Respiratory and Circulatory Systems

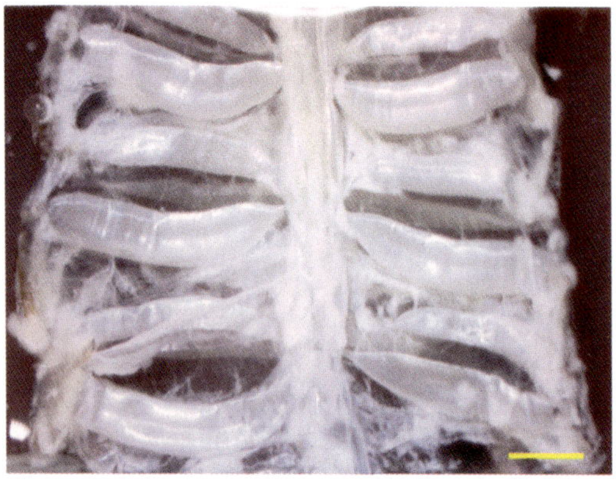

Tracheal system of a cockroach. The largest tracheae run across the width of the body and are horizontal in this image. Scale bar: 2 mm

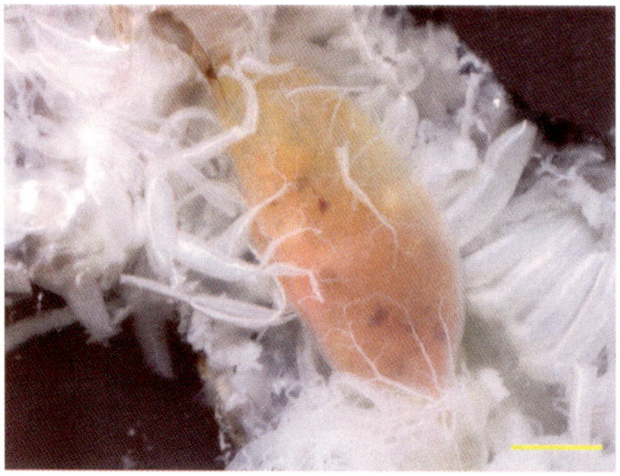

The tracheal system branches into ever smaller tubes. here they supply the crop of the cockroach. Scale bar: 2 mm

Insect respiration happens without lungs. There is a system of internal tubes and sacs through which gases diffuse or are actively pumped. Air is taken in through openings on the sides of the abdomen called spiracles. Oxygen gets to tissues that need it through their trachea (element 8 in diagram).

Many insect larvae live in water. Many of those have gills that can extract oxygen dissolved in water. Others must rise to the water surface to get air which may be held or trapped in special parts of their body.

Adult insects use oxygen at a high rate when they fly. They need it for the flight muscles, the most active tissue known in biology. The flight muscles use oxygen at a huge rate: 100 ccs of oxygen for every single cc of tissue per hour. With this system, the greatest diameter a muscle could have (and still consume oxygen at this rate) is about 0.5cm. Even with special extra arrangements, insects cannot get larger than about 11cm long. The largest insect bodies are about as big as a mouse.

Some insects also use a molecule called haemocyanin, which does the same job as haemoglobin does in vertebrates (but less efficiently). The insect circulatory system has no veins or arteries. The 'blood' is called haemolymph, and moves around in the space called the haemocoel. The organs sit in the haemocoel and are bathed in the haemolymph. The 'heart' is little more than a single tube which pulses (squeezes).

How Insects Grow

Insects start life as an egg. Usually a female (mother) insect lays eggs, but a few species have live birth (the eggs develop inside the mother). The eggs are small; but they can usually be seen with the naked eye.

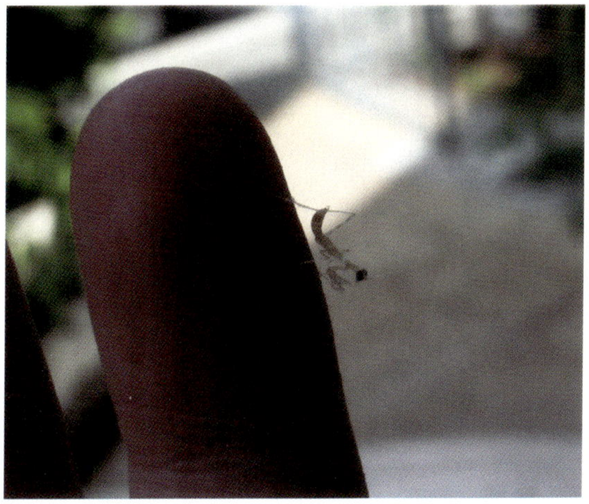

A mantis nymph looks just like a mantis adult but much smaller.

Although the adults are larger, they do need a magnifying glass or a binocular microscope to see the details. A professional entomologist uses a binocular microscope to identify insects, plus a printed reference work. There are far too many insects for anyone to remember them all, and most entomologists specialise in just one or two orders.

After the eggs hatch, two kinds of development may occur. Some insects have what is called 'incomplete metamorphosis'. This means that a small insect, called a nymph comes out of the egg, and the nymph looks almost the same as the adult insect. As the nymph grows, it does not change the way it looks, but only how big it is. It goes through a number of stages, called 'instars'. Grasshoppers grow in this way.

Other insects have complete metamorphosis, which means that the small larva which comes out of the egg looks very different from the adult insect. Insects that have complete metamorphosis usually come out of the egg as a larva, which usually looks like a worm. The larva eats food and gets bigger until it turns into a pupa. Butterfly pupae (plural for pupa) are often inside cocoons. Inside the cocoon the insect changes the way it looks and often grows wings. When the cocoon opens, the adult insect comes out. Many insects have complete metamorphosis, for example beetles, butterflies and moths, and flies. The adult stage of development is called the imago.

Evolutionary History

Origin of Insects

The oldest known insect fossil is the Devonian *Rhyniognatha*, from the 396 million year old Rhynie chert. It may have superficially resembled a modern-day silverfish insect. This species already possessed mandibles of a type associated with winged insects, suggesting that wings may already have evolved at this time. Thus, the first insects probably appeared earlier, in the Silurian period.

Origin of Wings

In 2008, researchers uncovered what they believe is the world's oldest known full-body impression of a primitive flying insect, a 300 million-year-old specimen from the Carboniferous period.

The origins of insect flight remain obscure, since the earliest winged insects currently known appear to have been capable fliers. Some extinct insects had an additional pair of winglets attaching to the first segment of the thorax, for a total of three pairs. It seems the insects were not a particularly successful group of animals before they evolved wings.

Upper Carboniferous and Lower Permian insect orders include both living groups and a number of Palaeozoic groups, now extinct. During this era, some giant dragonfly-like forms reached wingspans of 55 to 70 cm (22 to 28 in) making them far larger than any living insect.

This gigantism may have been due to higher atmospheric oxygen levels, which allowed increased respiratory efficiency. The lack of flying vertebrates could have been another factor. Many of the early groups became extinct during the Permian-Triassic extinction event, the largest mass extinction in the history of the Earth, around 252 million years ago.

Kinds of Insects

A beetle (ladybird or ladybug). The red part is the hard front pair of wings, or elytra.

Different kinds of insects can be divided into groups called orders. There are many insect orders. The biggest insect orders are listed below:

- Beetles (order Coleoptera) have the front pair of wings changed into a hard shell to protect the back wings.
- Butterflies and moths (order Lepidoptera) have large, often colourful wings.
- Flies (order Diptera) have only two wings.
- Ants, bees, and wasps (order Hymenoptera) sometimes have stingers and sometimes live in large colonies (like ant hills).
- True bugs (order Hemiptera) have a mouth that is long and narrow, like a drinking straw. This kind of mouth is called a beak.

- Grasshoppers (order Orthoptera) can usually jump with their legs. Eat grass and grain plants.
- Odonata, dragonflies and damselflies are top predators of other insects. Both aquatic nymphs and flying adults are carnivorous.
- Phasmatodea, the stick and leaf insects, is an order which is entirely based on camouflage. It includes the world's longest insect, Chan's megastick.

Spiders, scorpions, and similar animals are not insects; they are arachnids. Arachnids are arthropods that have four pairs of legs. Centipedes are also arthropods, but not insects: they are in a subphylum called the Myriapoda.

Insects and People

Pests can be harmful to people in different ways. Some are parasites, such as lice and bed bugs. Some of these parasite insects can spread diseases, such as mosquitoes spreading malaria.

Other pest insects do not directly hurt people. Termites and some beetles eat wood and sometimes eat buildings, such as houses. However, termites also help break down the trees and branches that fall on the forest floor. Many insects eat agricultural products (plants meant for people to eat). Grasshoppers are one example of pest insects that eat plants in agriculture.

Some insects are useful to us. Bees make honey. The larvae of some moths make silk, which people use to make clothing. In some parts of the world, people actually eat insects. Eating insects for food is called entomophagy.

Many bees and flies pollinate plants. This means the insects help the plants make seeds by moving pollen from one flower to another. Some good insects eat pest insects, such as lady beetles (or ladybirds or ladybugs) eating aphids. Many insects eat dead plants and animals.

Pesticides

People often use poisons called insecticides to kill pest insects. Insecticides do not always work. Sometimes the pest insects become resistant to the insecticides, which means the insecticides do not hurt them anymore. Both the Colorado potato beetle and the diamondback moth are insects that are resistant to many insecticides.

Insecticides do not only kill pest insects; sometimes many helpful insects are killed too. When helpful insects are killed, such as those that eat pest insects, the pest insects may come back in larger numbers than before because they are not being eaten by helpful insects anymore.

Carcinology

Carcinology is a branch of zoology that deals with insectss. It is particularly one of the sub-fields of arthropodology. Arthropodology, in particular, is the biological science that studies arthropods. Arthropods are invertebrates (of the phylum Arthropoda) that are characterized mainly by possessing a hard chitinous exoskeleton and multiple paired jointed limbs. Since arthropodology involves the largest phylum of the animal kingdom it is further subdivided into these sub-disciplines:

arachnology (the study of arachnids), entomology (the study of insects), carcinology (the study of crustaceans), and myriapodology (the study of myriapods).

Carcinology is primarily concerned with the studying of crustaceans. It attempts to study and understand various biological aspects of crustaceans, such as development, taxonomy, anatomy, physiology, genetics, evolution, distribution, and ecology of crustaceans. It has taxon-based subdivisions. An example is astacology. It is the study of crayfish. Crayfish are freshwater crustaceans characterized by their feather-like gills and morphological features similar to a lobster. Another sub-discipline is cirripedology, which is the study of barnacles. Copepodology is the sub-discipline that studies copepods.

An expert in this field is called a *carcinologist*. The carcinologist performs studies or research that are focused at understanding the biology of crustaceans as well as their identification and classification, evolutionary relationships, ecological relationships, and distribution.

Crustacean

Crustaceans form a large, diverse arthropod taxon which includes such familiar animals as crabs, lobsters, crayfish, shrimp, krill, woodlice, and barnacles. The crustacean group is usually treated as a subphylum, and thanks to recent molecular studies it is now well accepted that the crustacean group is paraphyletic, and comprises all animals in the Pancrustacea clade other than hexapods. Some crustaceans are more closely related to insects and other hexapods than they are to certain other crustaceans.

The 67,000 described species range in size from *Stygotantulus stocki* at 0.1 mm (0.004 in), to the Japanese spider crab with a leg span of up to 3.8 m (12.5 ft) and a mass of 20 kg (44 lb). Like other arthropods, crustaceans have an exoskeleton, which they moult to grow. They are distinguished from other groups of arthropods, such as insects, myriapods and chelicerates, by the possession of biramous (two-parted) limbs, and by their larval forms, such as the nauplius stage of brachiopods and copepods.

Most crustaceans are free-living aquatic animals, but some are terrestrial (e.g. woodlice), some are parasitic (e.g. Rhizocephala, fish lice, tongue worms) and some are sessile (e.g. barnacles). The group has an extensive fossil record, reaching back to the Cambrian, and includes living fossils such as *Triops cancriformis*, which has existed apparently unchanged since the Triassic period. More than 10 million tons of crustaceans are produced by fishery or farming for human consumption, the majority of it being shrimp and prawns. Krill and copepods are not as widely fished, but may be the animals with the greatest biomass on the planet, and form a vital part of the food chain. The scientific study of crustaceans is known as carcinology (alternatively, *malacostracology*, *crustaceology* or *crustalogy*), and a scientist who works in carcinology is a carcinologist.

Structure

The body of a crustacean is composed of segments, which are grouped into three regions: the *cephalon* or head, the thorax, and the *pleon* or abdomen. The head and thorax may be fused together to form a cephalothorax, which may be covered by a single large carapace. The crustacean body is protected by the hard exoskeleton, which must be moulted for the animal to grow. The shell around each somite can be divided into a dorsal tergum, ventral sternum and a lateral pleuron. Various parts of the exoskeleton may be fused together.

A shed carapace of a lady crab, part of the hard exoskeleton

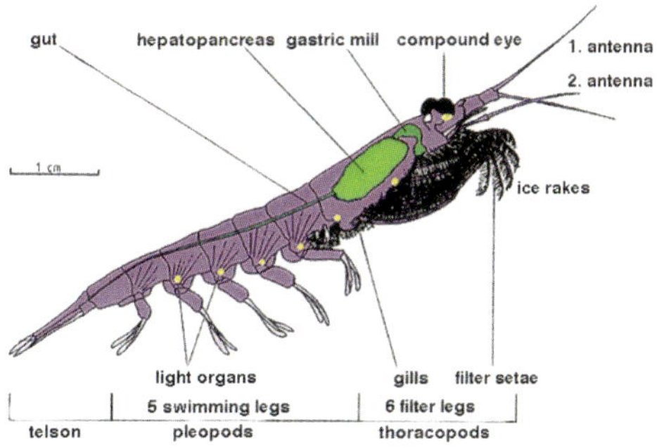

Body structure of a typical crustacean – krill

Each somite, or body segment can bear a pair of appendages: on the segments of the head, these include two pairs of antennae, the mandibles and maxillae; the thoracic segments bear legs, which may be specialised as pereiopods (walking legs) and maxillipeds (feeding legs). The abdomen bears pleopods, and ends in a telson, which bears the anus, and is often flanked by uropods to form a tail fan. The number and variety of appendages in different crustaceans may be partly responsible for the group's success.

Crustacean appendages are typically biramous, meaning they are divided into two parts; this includes the second pair of antennae, but not the first, which is usually uniramous, the exception being in the Class Malacostraca where the antennules may be generally biramous or even triramous. It is unclear whether the biramous condition is a derived state which evolved in crustaceans, or whether the second branch of the limb has been lost in all other groups. Trilobites, for instance, also possessed biramous appendages.

The main body cavity is an open circulatory system, where blood is pumped into the haemocoel by a heart located near the dorsum. Malacostraca have haemocyanin as the oxygen-carrying pigment, while copepods, ostracods, barnacles and branchiopods have haemoglobins. The alimentary canal consists of a straight tube that often has a gizzard-like "gastric mill" for grinding food and a pair of digestive glands that absorb food; this structure goes in a spiral format. Structures that function as kidneys are located near the antennae. A brain exists in the form of ganglia close to the antennae, and a collection of major ganglia is found below the gut.

In many decapods, the first (and sometimes the second) pair of pleopods are specialised in the male for sperm transfer. Many terrestrial crustaceans (such as the Christmas Island red crab) mate seasonally and return to the sea to release the eggs. Others, such as woodlice, lay their eggs on land, albeit in damp conditions. In most decapods, the females retain the eggs until they hatch into free-swimming larvae.

Ecology

The majority of crustaceans are aquatic, living in either marine or freshwater environments, but a few groups have adapted to life on land, such as terrestrial crabs, terrestrial hermit crabs, and woodlice. Marine crustaceans are as ubiquitous in the oceans as insects are on land. The majority of crustaceans are also motile, moving about independently, although a few taxonomic units are parasitic and live attached to their hosts (including sea lice, fish lice, whale lice, tongue worms, and *Cymothoa exigua*, all of which may be referred to as "crustacean lice"), and adult barnacles live a sessile life – they are attached headfirst to the substrate and cannot move independently. Some branchiurans are able to withstand rapid changes of salinity and will also switch hosts from marine to non-marine species. Krill are the bottom layer and the most important part of the food chain in Antarctic animal communities. Some crustaceans are significant invasive species, such as the Chinese mitten crab, *Eriocheir sinensis*, and the Asian shore crab, *Hemigrapsus sanguineus*.

Life Cycle

Eggs of *Potamon fluviatile*, a freshwater crab

Zoea larva of the European lobster, *Homarus gammarus*

Mating System

The majority of crustaceans have separate sexes, and reproduce sexually. A small number are hermaphrodites, including barnacles, remipedes, and Cephalocarida. Some may even change sex during the course of their life. Parthenogenesis is also widespread among crustaceans, where viable eggs are produced by a female without needing fertilisation by a male. This occurs in many branchiopods, some ostracods, some isopods, and certain "higher" crustaceans, such as the *Marmorkrebs* crayfish.

Eggs

In many groups of crustaceans, the fertilised eggs are simply released into the water column, while others have developed a number of mechanisms for holding on to the eggs until they are ready to hatch. Most decapods carry the eggs attached to the pleopods, while peracarids, notostracans, anostracans, and many isopods form a brood pouch from the carapace and thoracic limbs. Female Branchiura do not carry eggs in external ovisacs but attach them in rows to rocks and other objects. Most leptostracans and krill carry the eggs between their thoracic limbs; some copepods carry their eggs in special thin-walled sacs, while others have them attached together in long, tangled strings.

Larvae

Crustaceans exhibit a number of larval forms, of which the earliest and most characteristic is the nauplius. This has three pairs of appendages, all emerging from the young animal's head, and a single naupliar eye. In most groups, there are further larval stages, including the zoea. This name was given to it when naturalists believed it to be a separate species. It follows the nauplius stage and precedes the post-larva. Zoea larvae swim with their thoracic appendages, as opposed to nauplii, which use cephalic appendages, and megalopa, which use abdominal appendages for swimming. It often has spikes on its carapace, which may assist these small organisms in maintaining directional swimming. In many decapods, due to their accelerated development, the zoea is the first larval stage. In some cases, the zoea stage is followed by the mysis stage, and in others, by the megalopa stage, depending on the crustacean group involved.

Classification

The name "crustacean" dates from the earliest works to describe the animals, including those of Pierre Belon and Guillaume Rondelet, but the name was not used by some later authors, including Carl Linnaeus, who included crustaceans among the "Aptera" in his *Systema Naturae*. The earliest nomenclaturally valid work to use the name "Crustacea" was Morten Thrane Brünnich's *Zoologiæ Fundamenta* in 1772, although he also included chelicerates in the group.

The subphylum Crustacea comprises almost 67,000 described species, which is thought to be just $^1/_{10}$ to $^1/_{100}$ of the total number as the majority of species remain as yet undiscovered. Although most crustaceans are small, their morphology varies greatly and includes both the largest arthropod in the world – the Japanese spider crab with a leg span of 3.7 metres (12 ft) – and the smallest, the 1-micrometre-long (0.00004 in) *Stygotantulus stocki*. Despite their diversity of form, crustaceans are united by the special larval form known as the nauplius.

The exact relationships of the Crustacea to other taxa are not completely settled as of April 2012. Studies based on morphology led to the Pancrustacea hypothesis, in which Crustacea and Hexapoda (insects and allies) are sister groups. More recent studies using DNA sequences suggest that Crustacea is paraphyletic, with the hexapods nested within a larger Pancrustacea clade.

Although the classification of crustaceans has been quite variable, the system used by Martin and Davis largely supersedes earlier works. Mystacocarida and Branchiura, here treated as part of Maxillopoda, are sometimes treated as their own classes. Six classes are usually recognised:

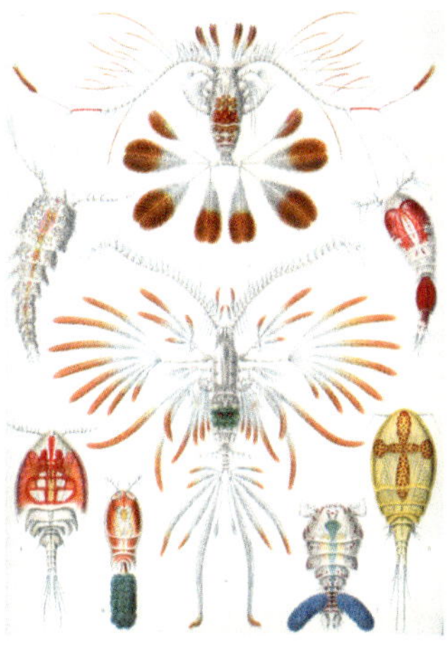

Copepods, from Ernst Haeckel's work *Kunstformen der Natur*

Decapods, from Ernst Haeckel's work *Kunstformen der Natur*

Class	Members	Orders	Photo
Branchiopoda	brine shrimp fairy shrimp water fleas tadpole shrimp clam shrimp	Anostraca Lipostraca Notostraca Laevicaudata Spinicaudata Cyclestherida Cladocera	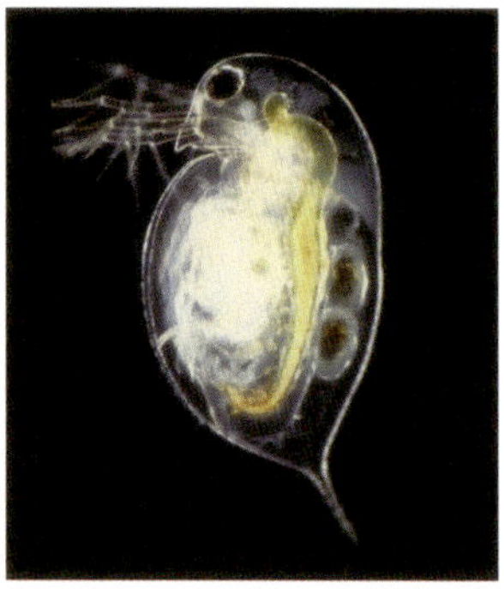 *Daphnia pulex* (Cladocera)
Remipedia		Nectiopoda	 *Speleonectes tanumekes* (Speleonectidae)
Maxillopoda	barnacles copepods	Calanoida Pedunculata Sessilia *c.* 20 others	 *Chthamalus stellatus* (Sessilia)
Ostracoda	seed shrimp	Myodocopida Halocyprida Platycopida Podocopida	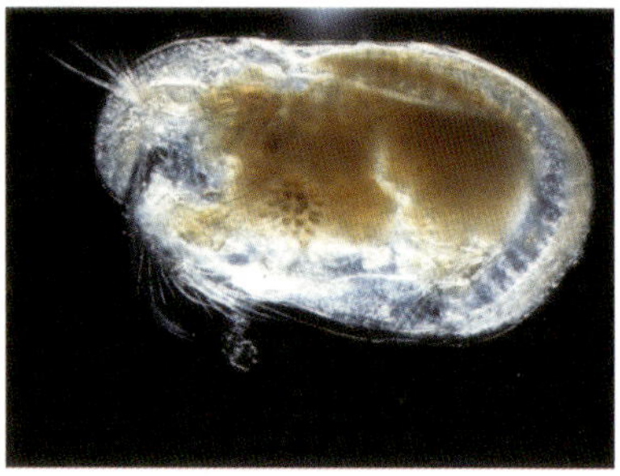 Cylindroleberididae
Malacostraca	crabs lobsters crayfish shrimp krill mantis shrimp woodlice hooded shrimp scuds sandhoppers *etc.*	Decapoda Isopoda Amphipoda Stomatopoda *c.* 12 others	 *Gammarus roeseli* (Amphipoda)

Fossil Record

Eryma mandelslohi, a fossil decapod from the Jurassic of Bissingen an der Teck, Germany

Crustaceans have a rich and extensive fossil record, which begins with animals such as *Canadaspis* and *Perspicaris* from the Middle Cambrian age Burgess Shale. Most of the major groups of crustaceans appear in the fossil record before the end of the Cambrian, namely the Branchiopoda, Maxillopoda (including barnacles and tongue worms) and Malacostraca; there is some debate as to whether or not Cambrian animals assigned to Ostracoda are truly ostracods, which would otherwise start in the Ordovician. The only classes to appear later are the Cephalocarida, which have no fossil record, and the Remipedia, which were first described from the fossil *Tesnusocaris goldichi*, but do not appear until the Carboniferous. Most of the early crustaceans are rare, but fossil crustaceans become abundant from the Carboniferous period onwards.

Norway lobsters on sale at a Spanish market

Within the Malacostraca, no fossils are known for krill, while both Hoplocarida and Phyllopoda contain important groups that are now extinct as well as extant members (Hoplocarida: mantis shrimp are extant, while Aeschronectida are extinct; Phyllopoda: Canadaspidida are extinct, while Leptostraca are extant). Cumacea and Isopoda are both known from the Carboniferous, as are the first true mantis shrimp. In the Decapoda, prawns and polychelids appear in the Triassic, and shrimp and crabs appear in the Jurassic; however, the great radiation of crustaceans occurred in the Cretaceous, particularly in crabs, and may have been driven by the adaptive radiation of their main predators, bony fish. The first true lobsters also appear in the Cretaceous.

Consumption by Humans

Many crustaceans are consumed by humans, and nearly 10,700,000 tons were produced in 2007; the vast majority of this output is of decapod crustaceans: crabs, lobsters, shrimp, crawfish, and prawns. Over 60% by weight of all crustaceans caught for consumption are shrimp and prawns, and nearly 80% is produced in Asia, with China alone producing nearly half the world's total. Non-decapod crustaceans are not widely consumed, with only 118,000 tons of krill being caught, despite krill having one of the greatest biomasses on the planet.

Myriapodology

Myriapodology is a branch of zoology that deals with myriapods. It is one of the sub-fields of arthropodology. Arthropodology, in particular, is the biological science that studies arthropods. Arthropods are invertebrates (of the phylum Arthropoda) that are characterized mainly by possessing a hard chitinous exoskeleton and multiple paired jointed limbs. Since arthropodology involves the largest phylum of the animal kingdom it is further subdivided into these sub-disciplines: arachnology (the study of arachnids), entomology (the study of insects), carcinology (the study of crustaceans), and myriapodology (the study of myriapods).

Myriapodology deals primarily with the myriapods with the intent of understanding the myriapods' development, taxonomy, biology, evolution, distribution, and ecology. Myriapods include the millipedes, the centipedes, the symphylans, the pauropodans, and the arthropleurideans.

An expert in this field is called a *myriapodologist*. The myriapodologists conduct studies in order to understand the biology of crustaceans as well as their identification and classification, evolutionary relationships, ecological importance, and distribution. One of the notable people that contributed in this field was Carl August Graf Attems-Petzenstein. He was an Austrian zoologist and myriapodologist who published 138 scientific papers, mainly on myriapods. He was able to describe about 1800 species and subspecies of myriapods from different parts of the world. Another was Henry Wilfred Brolemann, a French myriapodologist who named some 500 myriapod species.

Myriapoda

Myriapoda is a subphylum of arthropods containing millipedes, centipedes, and others. The group contains over 16,000 species, most of which are terrestrial. Although their name suggests they have myriad (10,000) legs, myriapods range from having up to 750 legs (the millipede *Illacme plenipes*) to having fewer than ten legs.

The fossil record of myriapods reaches back into the late Silurian, although molecular evidence suggests a diversification in the Cambrian Period, and Cambrian fossils exist which resemble myriapods. The oldest unequivocal myriapod fossil is of the millipede *Pneumodesmus newmani*, from the late Silurian (428 million years ago). *P. newmani* is also important as the earliest known terrestrial animal. The phylogenetic classification of myriapods is still debated.

The scientific study of myriapods is myriapodology, and those who study myriapods are myriapodologists.

Anatomy

Myriapods have a single pair of antennae and, in most cases, simple eyes. Exceptions include the large and well-developed compound eyes of Scutigera The mouthparts lie on the underside of the head, with an "epistome" and labrum forming the upper lip, and a pair of maxillae forming the lower lip. A pair of mandibles lie inside the mouth. Myriapods breathe through spiracles that connect to a tracheal system similar to that of insects. There is a long tubular heart that extends through much of the body, but usually few, if any, blood vessels.

The head of *Scutigera coleoptrata*, showing antennae, compound eyes and mouthparts

Malpighian tubules excrete nitrogenous waste into the digestive system, which typically consists of a simple tube. Although the ventral nerve cord has a ganglion in each segment, the brain is relatively poorly developed.

During mating, male myriapods produce a packet of sperm, or spermatophore, which they must transfer to the female externally; this process is often complex and highly developed. The female lays eggs which hatch as much-shortened versions of the adults, with only a few segments and as few as three pairs of legs. The young add additional segments and limbs as they repeatedly moult to reach the adult form.

Ecology

Myriapods are most abundant in moist forests, where they fulfill an important role in breaking down decaying plant material, although a few live in grasslands, semi-arid habitats or even deserts. A very small percentage of species are littoral (found along the sea shore). The majority are detritivorous, with the exception of centipedes, which are chiefly nocturnal predators. A few species of centipedes and millipedes are able to produce light and are therefore bioluminescent Pauropodans and symphylans are small, sometimes microscopic animals that resemble centipedes superficially and live in soils. Millipedes differ from the other groups in having their body segments fused into pairs, giving the appearance that each segment bears two pairs of legs, while the other three groups have a single pair of legs on each body segment.

Although not generally considered dangerous to humans, many millipedes produce noxious secretions (often containing benzoquinones) which in rare cases can cause temporary blistering and discolouration of the skin. Large centipedes, however, can bite humans, and although the bite may cause intense pain and discomfort, fatalities are extremely rare.

Classification

There has been much debate as to which arthropod group is most closely related to the Myriapoda. Under the Mandibulata hypothesis, Myriapoda is the sister taxon to Pancrustacea, a group comprising the Crustacea and Hexapoda (insects and their close relatives). Under the Atelocerata hypothesis, Hexapoda is the closest, whereas under the Paradoxopoda hypothesis, Chelicerata is the closest. This last hypothesis, although supported by few, if any, morphological characters, is supported by a number of molecular studies. There are four classes of extant myriapods, Chilopoda (centipedes), Diplopoda, Pauropoda and Symphyla, containing a total of around 12,000 species. While each of these groups of myriapods is believed to be monophyletic, relationships among them are less certain.

Centipedes

Scolopendra cingulata, a centipede

Centipedes make up the class Chilopoda. They are fast, predatory and venomous, hunting mostly at night. There are around 3,300 species, ranging from the diminutive *Nannarrup hoffmani* (less than 12 mm or ½ in in length) to the giant *Scolopendra gigantea*, which may exceed 30 centimetres (12 in).

Millipedes

Tachypodoiulus niger, a millipede

Millipedes form the class Diplopoda. Most millipedes are slower than centipedes, and feed on leaf litter and detritus. They are distinguished by the fusion of each pair of body segments into a single unit, giving the appearance of having two pairs of legs per segment. Around 12,000 species have been described, which may represent less than a tenth of the true global millipede diversity. The name "millipede" is a compound word formed from the Latin roots *millia* ("thousand") and *pes* (gen. *pedis*) ("foot"), although millipedes typically have between 36 and 400 legs. One species, *Illacme plenipes*, has the greatest number of legs of any animal, with 750. Pill millipedes are much shorter, and are capable of rolling up into a ball, like pillbugs.

Symphyla

About 200 species of them are known worldwide. They resemble centipedes but are smaller and translucent. Many spend their lives as soil infauna, but some live arboreally. Juveniles have six pairs of legs, but, over a lifetime of several years, add an additional pair at each moult so that the adult instar has twelve pairs of legs.

Scutigerella immaculata, a symphylan

Pauropoda

Pauropoda is another small group of small myriapods. They are typically 0.5–2.0 mm long and live in the soil on all continents except Antarctica. Over 700 species have been described. They are believed to be the sister group to millipedes, and have the dorsal tergites fused across pairs of segments, similar to the more complete fusion of segments seen in millipedes.

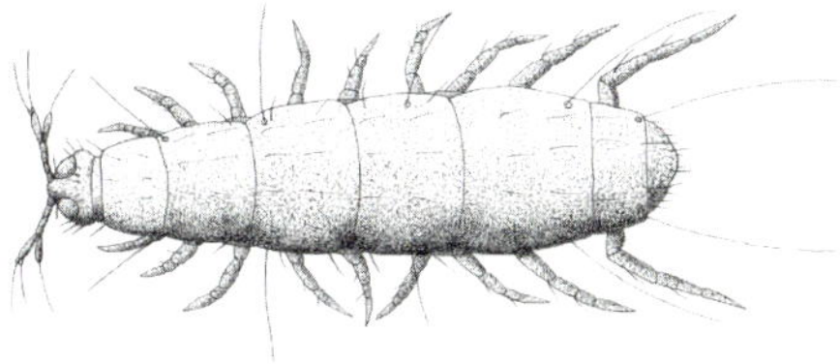

Pauropus huyxleyi, a pauropodan

Arthropleuridea

Arthropleurideans were ancient myriapods that are now extinct, known from the late Silurian to the Permian. The most famous members are from the genus *Arthropleura*, which was a giant, probably herbivorous, animal that could be up to 3 metres (10 ft) long, but the group also includes species less than 1 cm (0.39 in). Arthropleuridea was historically considered a distinct class of myriapods, but since 2000 scientific consensus has viewed the group as a subset of millipedes, although the relationship of arthropleurideans to other millipedes and to each other is debated.

Myriapod Relationships

A variety of groupings (clades) of the myriapod classes have been proposed, and married, some of which are mutually exclusive, and all of which represent hypotheses of evolutionary relationships. Traditional relationships supported by morphological similarities (anatomical or developmental similarities) are challenged by newer relationships supported by molecular evidence (including DNA sequence and amino acid similarities).

- Dignatha (also called Collifera) is a clade consisting of millipedes and pauropods, and is supported by morphological similarities including the presence of a gnathochilarium (a modified jaw and plate apparatus) and a collum, a legless segment behind the head.

- Trignatha (also called Atelopoda) is a grouping of centipedes and symphylans, united by similarities of mouthparts.

- Progoneata is a group encompassing Dignatha and symphylans while excluding centipedes. Shared features include reproductive openings (gonopores) behind the second body segment, and sensory hairs (trichobothria) with a bulb-like swelling.
- Edafopoda is a grouping of symphylans and pauropodans that is supported by shared genetic sequences, yet conflicts with Dignatha and Trignatha.

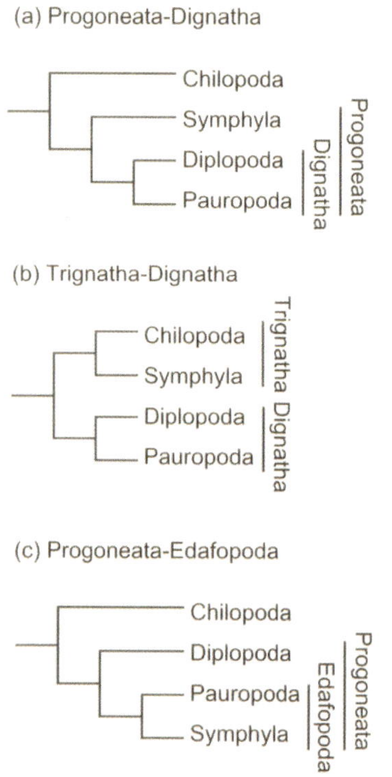

Some of the various hypotheses of myriapod phylogeny. Morphological studies (trees a and b) support a sister grouping of Diplopoda and Pauropoda, while studies of DNA or amino acid similarities suggest a variety of different relationships, including the relationship of Pauropoda and Symphyla in tree c.

Cnidariology

Cnidariology is a sub-discipline of invertebrate zoology that studies the cnidarians. Invertebrate zoology is concerned with the invertebrates. Invertebrates are animals that lack a vertebral column (a backbone). Since there are a number of invertebrate species there are taxonomic-based sub-disciplines of invertebrate zoology, particularly arthropodology (the study of arthropods), cnidariology, and helminthology (the study of helminths).

Cnidariology deals with studying the cnidarians. Cnidarians are those belonging to phylum Cnidaria. They include the jellyfishes, sea anemones, corals, hydras, *etc*. They are characterized primarily by possessing cnidae, i.e. specialized cellular structures that fire off thread-like tubules producing a stinging effect.

An expert on cnidariology is called *cnidariologist*. A cnidariologist would be concerned and responsible in conducting research in cnidarians. The research may be associated with the identification and classification of cnidarian species. It may also be about their distribution, conservation, and ecological relationship of the cnidarians to other living things and their environment. The research may also encompass other related fields such as genetics, evolution, and medical importance of cnidarians. An example of a work in cnidariology is that of Alfred G. Mayer, an American marine zoologist who documented his studies on various species of jellyfish around the world through his book entitled *Medusae of the World* (1910).

Cnidaria

Cnidaria is a phylum containing over 9,000 species found only in aquatic and mostly marine environments.

All cnidarians have radial symmetrical. There are two major body forms among the Cnidaria - the polyp and the medusa. Sea anemones and corals have the polyp form, while jellyfish are typical medusae. When you look at them, you can see that these body forms are the same except that one is upside down.

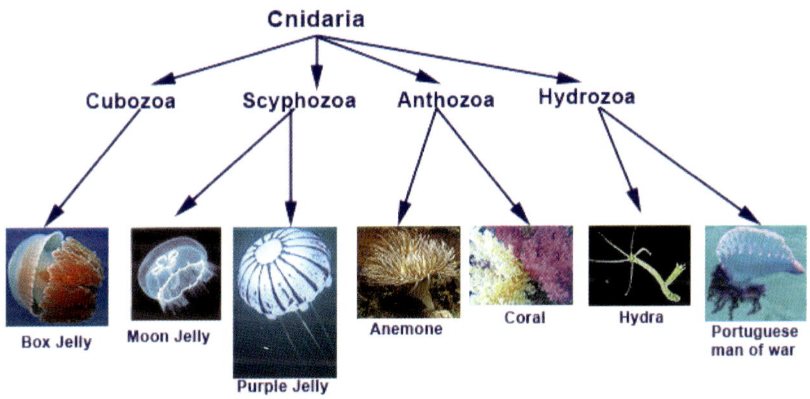

Cnidarians are classified into four main groups: non-moving (sessile) Anthozoa (sea anemones, corals, sea pens); Cubozoa (box jellies); swimming Scyphozoa (jellyfish); and Hydrozoa, a broad group that includes all the freshwater cnidarians as well as many marine forms. There are about 10, 000 species of Cnidarians.

Cnidaria have no organs like hearts or lungs. They have an internal cavity used for respiration and a gastrovascular cavity (a "stomach") with a mouth but they do not have an anus. They possess a body wall made from two layers (the ectoderm and the endoderm) separated by a jelly-like layer called the mesoglea. Other important functions like respiration and excretion occur by diffusion.

Diffusion is the movement of a material from regions of higher concentration to regions of lower concentration until equilibrium is reached. Diffusion is a natural process caused by the random movement of particles but is only useful to very small (or thin) organisms (ones which have a large surface area compared to their volume) where the material in higher concentration does not have far to move.

Cnidaria do not have a brain or groups of nerve cells ("ganglia"). The nervous system is a decentralized network ('nerve net'), with one or two nets present. They do not have a head, but they have a mouth, surrounded by a crown of tentacles. The tentacles are covered with stinging cells (nematocysts).

Body form of Cnidaria

Nearly all cnidaria are predators, their nematocysts can paralyze and kill prey much larger than them. Small sensory hairs near the nematocysts are sensitive to vibrations in the water. Any prey swimming past can trigger the nematocyst which shoots out the barb. This penetrates the prey's outer covering and injects it with venom. The prey is then moved to the mouth by a tentacle.

Their prey can range in size from plankton to animals several times larger than themselves. Some obtain their energy from algae that that live in their bodies and a few are parasites. Other Cnidaria, including the corals, get their nutrients from symbiotic algae within their cells.

Predators of Cnidaria include sea slugs, sea stars (for example, the Crown of Thorn which can devastate coral reefs), nudibranchs, fish including butterfly fish and parrot fish, which eat corals and marine turtles and sunfish, which eat jellyfish. Plastic bags floating in the oceans are dangerous to turtles which often can mistake them for jellyfish and their guts can become blocked by them.

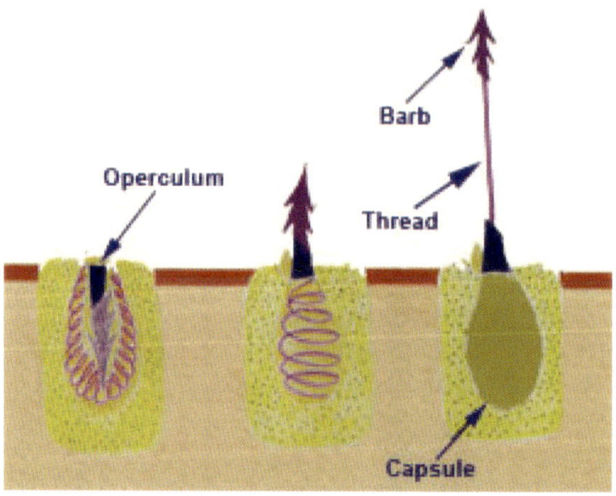

How a nematocyst works

For example coral, inside the sac of each coral polyp lives a singlee-celled algae called zooxanthellae. The algae produces oxygen and energy (sugars) that the coral polyp needs to live and, in return, the polyp produces carbon dioxide and other substances the algae needs. That is why coral reefs grow so near the surface of the water where the algae can get sunshine for photosynthesis.

Many Cnidaria live in colonies made up of large numbers of individuals joined together in some way. These individuals (called zooids) can either be directly connected by tissues or share a common exoskeleton made from chiton or calcium carbonate.

In the Cnidaria sexual reproduction often involves a complex life cycle with both polyp and medusa stages. For example in Scyphozoa (jellyfish) and Cubozoa (box jellies) a larva swims until it finds a good site, and then becomes a polyp. This grows normally but then absorbs its tentacles and splits horizontally into a series of disks that become juvenile medusae, a process called strobilation. The juveniles swim off and slowly grow to maturity, while the polyp re-grows and may continue strobilating periodically. The adults have gonads in the gastroderm, and these release ova and sperm into the water in the breeding season.

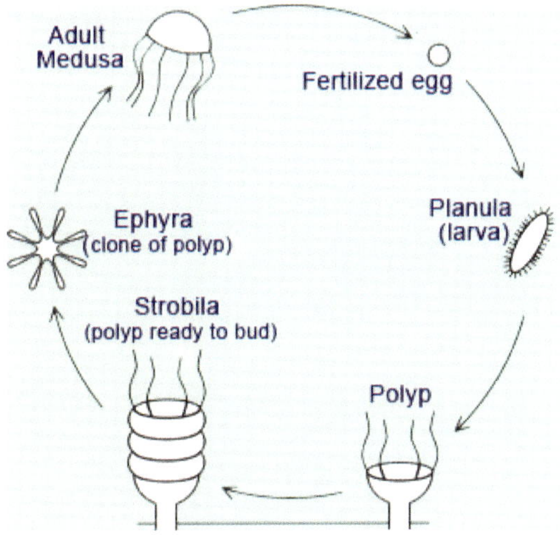

Life cycle of a jellyfish

Other Cnidaria have shorter forms of this cycle, for example cubozoan polyps have only one medusa stage. All Cnidaria can reproduce asexually by various means, in addition to regenerating if their bodies are divided into segments or are attacked by predators. Some produce buds, others divide down the middle, others do both.

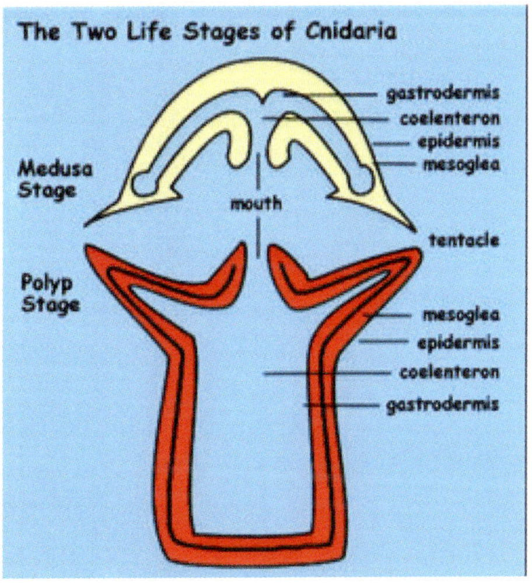

Structure of Cnidaria
From Life on Australian Seashores

Cnidarians range in size from Hydra, 5–20 millimetres to the Lion's mane jellyfish, which may exceed 2 metres in diameter and 75 metres in length.

Characteristics of Cnidarians

Cnidarians exhibit radial symmetry, which means their body parts are arranged symmetrically around a central axis. So, if you drew a line from any point at the edge of a cnidarian through the center and to the other side, you'd have two roughly equal halves.

Cnidarians also have tentacles. These tentacles have stinging structures called cnidocytes, which bear nematocysts. Cnidarians got their name from these stinging structures. The word cnidarian comes from the Greek word *knide* (nettle).

The presence of nematocysts is a key feature of cnidarians. Cnidarians can use their tentacles for defense or for capturing prey.

Although they can sting, not all cnidarians pose a threat to humans. Some, like the box jellyfish, have very potent toxins in their tentacles, but others, like moon jellies, have toxins that don't have enough power to sting us.

Cnidarians have two body layers called the epidermis and gastrodermis. Sandwiched in between is a jelly-like substance called mesoglea.

Examples of Cnidarians

As a large group comprised of thousands of species, cnidarians can be pretty diverse in their form. Overall, though, they have two main body plans: polypoid, in which the mouth faces up (e.g., anemones) and medusoid, in which the mouth faces down (e.g., jellyfish). Cnidarians may go through stages in their life cycle in which they experience each of these body plans.

There are several major groups of cnidarians:

- Anthozoa: sea anemones, sea pens, and corals. These animals have a polypoid body plan and attach to a substrate, such as other animals, rocks or algae.
- Hydrozoa: hydrozoans, also known as hydromedusae or hydroids. These organisms alternate between polyp and medusa stages and are usually colonial organisms. Siphonophores, which include Portuguese man-of-war and by-the-wind sailors, are examples of animals in the Class Hydrozoa. Most cnidarians are marine organisms, but there are some hydrozoan species that live in fresh water.
- Scyphozoa or Scyphomedusae: true jellyfish are in the Class Scyphozoa. These animals are known for their bell shape with dangling oral arms. Some jellyfish have tentacles also. The lion›s mane jellyfish is the largest species, with tentacles that may stretch more than 100 feet.
- Cubozoa: box jellyfish. These animals have a cube-shaped bell, with tentacles dangling from each corner. The sea wasp, a type of box jellyfish, is said to be the most venomous marine animal.
- Staurozoa: stalked jellyfish or Stauromedusae. These strange-looking, trumpet-shaped an-

imals aren't free-swimming like regular jellyfish. Instead, they attach to rocks or seaweed and are typically found in cold water.

- Myxozoa: parasitic microorganisms that evolved from jellyfish There has been debate over the years over where these animals should be classified - the latest research places them in the Cnidaria phylum, and an important piece of evidence is that these creatures have nematocysts. Myxozoa species can affect fish, worms, amphibians, reptiles and even mammals. One economic impact is that they can affect farmed fish such as salmon.

Smallest and Largest Cnidarians

The smallest cnidarian is a hydra with the scientific name *Psammohydra nanna*. This animal is less than half a millimeter in size.

The largest non-colonial cnidarian is the lion's mane jellyfish. As mentioned above, the tentacles are thought to stretch more than 100 feet. The bell of this jellyfish can be over 8 feet across.

Of colonial cnidarians, the longest is the giant siphonophore, which can grow to over 130 feet.

Helminthology

Helminthology is the study of parasitic worms (helminths), while helminthiasis describes the medical condition of being infected with helminths. Helminthology studies the taxonomy of helminths and their effects on their hosts.

The origin of the first compound of the word is the Greek helmins, meaning "worm".

In the 18th and early 19th century there was wave of publications on helminthology; this period has been described as the science's "Golden Era". During that period the authors Félix Dujardin, William Blaxland Benham, Peter Simon Pallas, Marcus Elieser Bloch, Otto Friedrich Müller, Johann Goeze, Friedrich Zenker, Charles Wardell Stiles, Carl Asmund Rudolphi, Otto Friedrich Bernhard von Linstow and Johann Gottfried Bremser started systematic scientific studies of the subject.

The helminth *Spinochordodes* parasitising a bush-cricket (*Meconema* sp.)

A plate from Félix Dujardin's

The Japanese parasitologist Satyu Yamaguti was one of the most active helminthologists of the 20th century; he wrote the six-volume *Systema Helminthum*.

Helminths

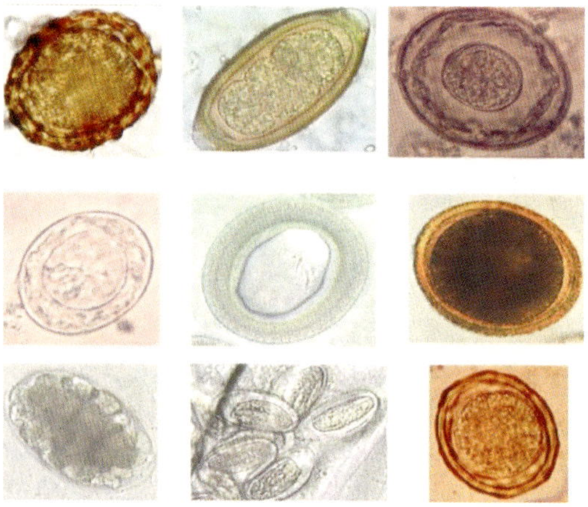

Helminth eggs of different helminth species, which are produced for reproduction of the worms.

Helminths also commonly known as parasitic worms, are large multicellular parasites, which can generally be seen with the naked eye when they are mature. They are often referred to as intestinal worms even though not all helminths reside in the intestines. For example, schistosomes are not intestinal worms, but rather reside in blood vessels. The word *helminth* comes from ancient Greek *hélmins*, an intestinal worm.

There is no consensus on the taxonomy of helminths. It is simply a commonly used term to describe certain worms with some similarities. These are flatworms (platyhelminthes), namely cestodes (tapeworms) and trematodes (flukes), and roundworms or nemathelminths (nematodes) – both of these are parasitic worm types – and the annelida, which are not parasitic or at the most ectoparasites like the leeches.

Helminths are worm-like organisms living in and feeding on living hosts. They receive nourishment and protection while disrupting their hosts' nutrient absorption. This can cause weakness and disease of the host. Those helminths that live inside the digestive tract are called intestinal parasites. They can live inside humans and other animals. In their adult form, helminths cannot multiply in humans. Helminths are able to survive in their mammalian hosts for many years due to their ability to manipulate the immune response by secreting immunomodulatory products. All helminths produce eggs (also called ova) for reproduction. These eggs have a strong shell that protects them against a range of environmental conditions. The eggs can therefore survive in the environment, outside their hosts, for many months or years.

Many, but not all, of the worms referred to as helminths belong to the group of intestinal parasites. An infection by a helminth is known as helminthiasis, helminth infection or intestinal worm infection. There is a naming convention which applies to all helminths: the ending "-asis" (or in veterinary science: "-osis") is added at the end of the name of the worm to denote the infection with that particular worm. For example, *Ascaris* is the name of a type of helminth, and ascariasis is the name of the infection caused by that helminth.

Taxonomy

There is no real consensus on the taxonomy (or groupings) of the helminths, particularly with the nematodes. The term "helminth" contains a number of phyla, many of which are completely unrelated. However, for practical considerations the term is still used nowadays to describe four groups with superficial similarities, the phyla Annelida, Platyhelminths, Nematoda and Acanthocephala.

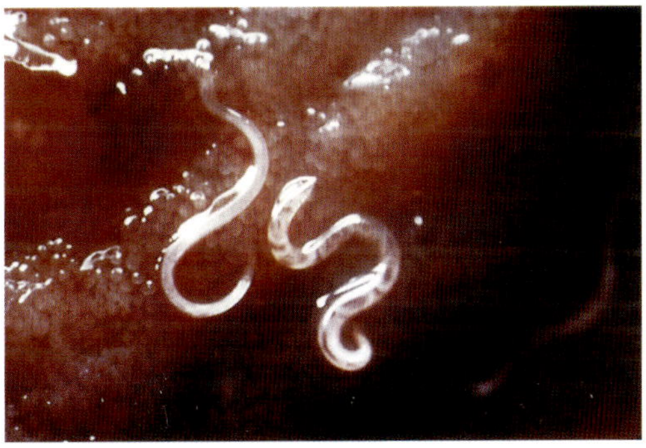

Hookworms attached to the intestinal mucosa

There is in fact no helminth classification; it is an "artificial" term.

The most important helminths in the sanitation field are the human parasites, which is why most people relate the term helminth to them, where they are classified as nemathelminthes (nematodes) and platyhelminthes, depending on whether they possess a round or flat-shaped body respectively. The latter are further divided into cestodes and trematodes depending on whether or not they have a segmented body.

Ringworm (dermatophytosis) is actually caused by various fungi and not by a parasitic worm.

Common Characteristics

Helminths are a group of organisms which share a similar form but are not necessarily related as part of evolution. Helminths include members of the following taxa: monogeneans, cestodes (tapeworms), nematodes (roundworms), and trematodes (flukes).

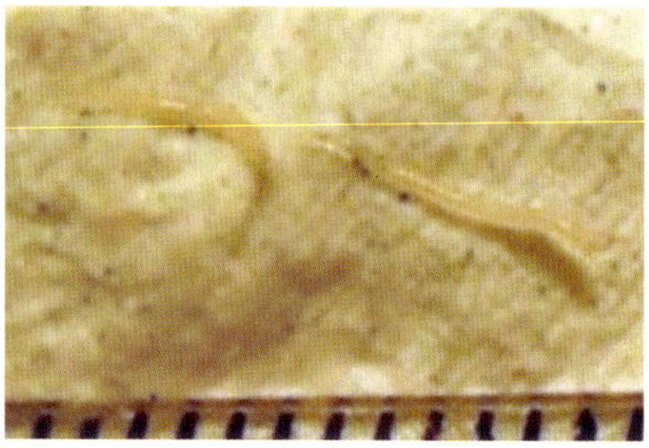

Two pinworms

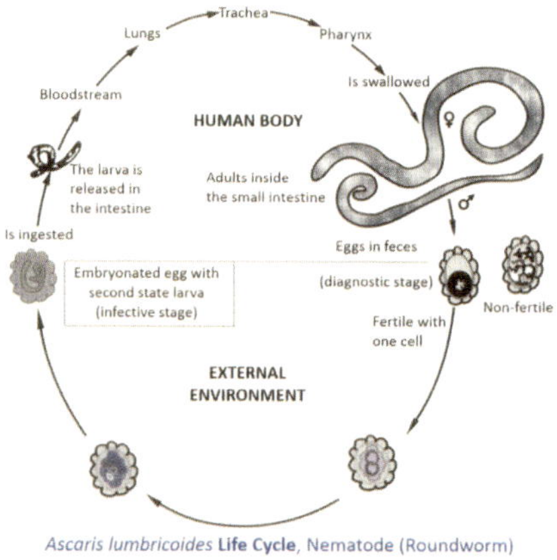

Image showing life cycle inside and outside of the human body of one fairly typica and well described helminth: Ascaris lumbricoides

The number of different helminth species is vast: it is estimated to be around one million species. The nematodes are the most diverse of all the helminths with the highest number of species. There may be as many as 300,000 species of parasites affecting vertebrates, and as many as 300 affecting humans alone.

Characteristics that are common for all helminths include:

Life time:

- The life time of adult worms varies tremendously from one species to another but is generally in the range of 1 to 8 years. This life time of several years is a result of their ability to manipulate the immune response of their hosts by secreting immunomodulatory products.
- Helminths can be either hermaphrodites (i.e. can have both sexes), like tapeworms and the flukes (except the blood fluke which is not a hermaphrodite), or have their sexes differentiated, like the roundworms.

Eggs:

- All helminths produce eggs (also called ova) for reproduction.
- Helminth eggs have a strong shell that protects them against a range of environmental conditions. This shell consists of three layers: a lipoidal inner layer, a chitinous middle layer and outer proteinic layer.

- Generally thousands or even hundreds of thousands of eggs are produced each time the female worm deposits its eggs - a process called ovoposition. The following table shows a large variation in the number of eggs produced by the different worms in one event; it varies in the range of 3,000 to 700,000.
- The frequency of egg deposition from an adult helminth is generally daily, or up to six times per day for some *Taenia* species.
- Adult trematodes lay smaller numbers of eggs compared to cestodes or nematodes. However, the egg develops into a miracidia from which thousands of *cercariae*, swimming larvae, develop. This means that one egg may produce thousands of adult worms.
- Helminth eggs remain viable for 1–2 months in crops and for many months in soil, fresh water and sewage, or even for several years in feces, fecal sludge (historically called night soil) and sewage sludge; a period that is much longer compared to other kinds of microorganisms.

Larvae:

- Larvae hatch from these eggs (if the eggs are viable), inside or outside the host, depending on the type of helminth. Life cycles of the helminths differ in this and other specific aspects. Eggs that are no longer viable do not produce any larvae.
- The larvae maturing in the host take from about two weeks up to four months depending on the helminth species.

Use in Medicine

Parasitic worms have been used as a medical treatment for various diseases, particularly those involving an overactive immune response. As humans have evolved with parasitic worms, proponents argue they are needed for a healthy immune system. Scientists are looking for a connection between the prevention and control of parasitic worms and the increase in allergies such as hay-fever in developed countries. Parasitic worms may be able to damp down the immune system of their host, making it easier for them to live in the intestine without coming under attack. This may be one mechanism for their proposed medicinal effect.

One study suggests a link between the rising rates of metabolic syndrome in the developed worlds and the largely successful efforts of Westerners to eliminate intestinal parasites. The work suggests eosinophils (a type of white blood cell) in fat tissue play an important role in preventing insulin resistance by secreting interleukin 4, which in turn switches macrophages into "alternative activation". Alternatively-activated macrophages are important to maintaining glucose homeostasis (i.e., blood sugar regulation). Helminth infection causes an increase in eosinophils. In the study, the authors fed rodents a high-fat diet to induce metabolic syndrome, and then injected them with helminths. Helminth infestation improved the rodents' metabolism. The authors concluded:

> Although sparse in blood of persons in developed countries, eosinophils are often elevated in individuals in rural developing countries where intestinal parasitism is prevalent and metabolic syndrome rare. We speculate that eosinophils may have evolved to optimize metabolic homeostasis during chronic infections by ubiquitous intestinal parasites....

Eggs

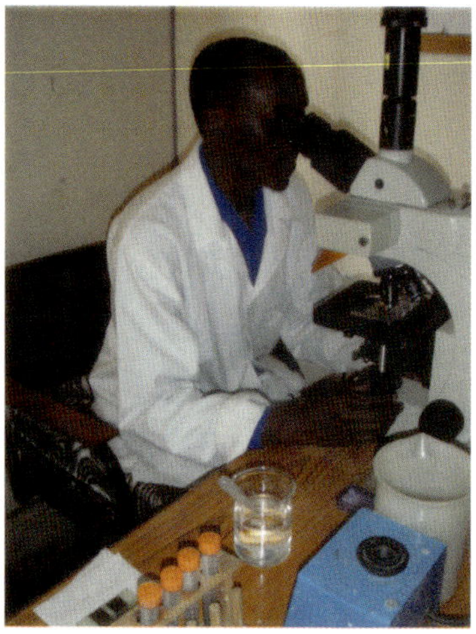

Analysing for helminth eggs in samples of feces from a dry toilet in Kenya

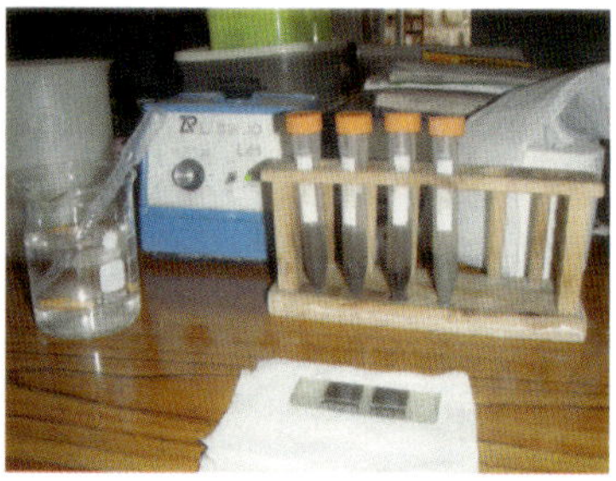

Processed helminth eggs samples from a dry toilet in Kenya

In the Environment

Eggs can reach the soil when polluted wastewater, sewage sludge or human excreta are used as fertilizer. Such soil is often characterized by moist and warm conditions. Therefore, the risk of using contaminated wastewater and sludge in agricultural fields is a real problem, especially in poor countries, where this practice is prevalent. Helminth eggs are regarded as the main biological health risk when applying sewage sludge, fecal sludge or fecal matter on agricultural soils. The eggs are the infective stage of the helminths' life cycle for causing the disease helminthiasis.

Helminth eggs are resistant to various environmental conditions due to the composition of the egg shell. Each helminth egg species has 3 to 4 layers with different physical and chemical characteristics: a) the 1 to 2 outer layers are formed of mucopolysaccharides and proteins, b) the middle layers consist of chitinous material and serve to give structure and mechanical resistance to the eggs, and c) the inner layer is composed of lipids and proteins and is useful to protect eggs from desiccation, strong acid and bases, oxidants and reductive agents as well as detergent and proteolytic compounds.

Due to this strong shell, helminth eggs or ova remain viable in soil, fresh water and sewage for many months. In feces, fecal sludge and sewage sludge they can even remain viable for several years. Helminth eggs of concern in wastewater used for irrigation have a size between 20 and 90 µm and a relative density of 1.06–1.23. It is very difficult to inactivate helminth eggs, unless temperature is increased above 40 °C or moisture is reduced to less than 5%.

Levels of Infectiousness

Helminth eggs contained in wastewater, sewage sludge or human excreta are not always infectious, i.e. able to cause the disease helminthiasis. Fertilized eggs and unfertilized eggs can exist side by side. Unfertilized eggs are identifiable under the microscope by their elongated shape. No larvae can hatch from these kinds of eggs. Therefore, unfertilized eggs do not pose a danger to human health.

In the case of *Ascaris lumbricoides* (giant roundworm), which has been considered the most resistant and common helminth type, fertilized eggs deposited in soil are resistant to desiccation but are, at this stage of development, very sensitive to environmental temperatures: The reproduction of a fertilized egg within the eggshell develops at an environmental soil temperature about 25 °C which is lower than the body temperature of the host (i.e., 37 °C for humans).

However, development of the larvae in the egg stops at temperatures below 15.5 °C, and eggs cannot survive temperatures much above 38 °C. If the temperature is around 25 °C, the infectiousness occurs after nearly 10 days of incubation. Finally, after 2 to 4 weeks in moist soil at optimal temperature and oxygen levels, the embryo develops into an infective larva, named "second-stage larva". This larva has the ability to get out of the egg, hatch in the small intestine and migrate to different organs. These infective larvae (or "infective eggs") may remain viable in soil for two years or longer.

Removal versus Inactivation in Wastewater and Sludge Treatment

In order to physically remove (but not inactivate) helminth eggs from wastewater, processes that remove particles, such as sedimentation, filtration or coagulation-flocculation are employed. Therefore, waste stabilization ponds (lagoons), storage bassins, constructed wetlands, rapid filtration or upflow anaerobic sludge blanket (UASB) reactors can be used. These conventional wastewater treatment processes do not inactivate the helminth ova but only removes them from the wastewater and moves them to the sewage sludge.

Helminth ova cannot be inactivated with chlorine, UV light or ozone (in the latter case at least not with economical doses because >36 mg/L ozone are needed with 1 hour contact time).

Inactivation of helminth ova can be achieved in sewage sludge treatment where the temperature is increased over 40 °C or moisture is reduced to less than 5%. Best results can be obtained when both of these conditions are combined for an extended period of time. Details about the contact time under these conditions and other related environmental factors are generally not well-defined for every type of helminth egg species. Helminth eggs are considered highly resistant biological structures.

Indicator Organism

Helminth eggs (or ova) are a good indicator organism to assess the safety of sanitation and reuse systems because they are the most environmentally resistant pathogens of all pathogens (viruses, bacteria, protozoa and helminths) and can in extreme cases survive for several years in soil. Therefore, the presence or absence of viable helminth eggs ("viable helminth egg" means that a larva would be able to hatch from the egg) in a sample of dried fecal matter, compost or fecal sludge is often used as an indicator to assess the efficiency of diverse wastewater and sludge treatment processes in terms of pathogen removal. In particular, the number of viable Ascaris eggs is often taken as an indicator organism for all helminth eggs in treatment processes as they are very common in many parts of the world and relatively easy to identify under the microscope. However, the exact inactivation characteristics may vary for different types of helminth eggs.

Measurement Techniques

Environmental Samples

For the purpose of setting treatment standards and reuse legislation, it is important to be able to determine the amount of helminth eggs in an environmental sample with some accuracy. The detection of viable helminth eggs in samples of wastewater, sludge or fresh feces (as a diagnostic tool for the infection helminthiasis) is not straight forward. In fact, many laboratories in developing countries lack the right equipment or skilled staff required to do so. An important step in the analytical methods is usually the concentration of the eggs in the sample, especially in the case of wastewater samples. A concentration step may not be required in samples of dried feces, e.g. samples collected from urine-diverting dry toilets.

Human Stool Samples

For medical purposes, the exact number of helminth eggs is less important and therefore most diagnoses are made simply by identifying the appearance of the worm or eggs in feces. Due to the large quantity of eggs laid, physicians can diagnose using only one or two fecal smears. The Kato technique (also called the Kato-Katz technique) is a laboratory method for preparing human stool samples prior to searching for parasite eggs. Eggs per gram is a laboratory test that determines the number of eggs per gram of feces in patients suspected of having a parasitological infection, such as schistosomiasis.

Malacology

Malacology is the study of mollusks. These animals include snails, slugs, clams, mussels, octopus, squid, and many other kinds of animals without backbones that are often found in the sea, but sometimes in lakes and rivers, and on land. Most of them have shells, but not all of them. Some people save these shells because they may be very pretty. Many people eat mollusks, especially those people who live near the ocean.

Scientists study mollusks to learn how many kinds there are, where they are found, how they came to be as they are, and what their lives are like, as well as many other things about them. These scientists are called *malacologists*.

Mollusca

Mollusca is a large phylum of invertebrate animals whose members are known as molluscs or mollusks. Around 85,000 extant species of molluscs are recognized. The number of fossil species is estimated between 60,000 and 100,000 additional species.

Molluscs are the largest marine phylum, comprising about 23% of all the named marine organisms. Numerous molluscs also live in freshwater and terrestrial habitats. They are highly diverse, not just in size and in anatomical structure, but also in behaviour and in habitat. The phylum is typically divided into 9 or 10 taxonomic classes, of which two are entirely extinct. Cephalopod molluscs, such as squid, cuttlefish and octopus, are among the most neurologically advanced of all invertebrates—and either the giant squid or the colossal squid is the largest known invertebrate species. The gastropods (snails and slugs) are by far the most numerous molluscs in terms of classified species, and account for 80% of the total.

Cornu aspersum (formerly *Helix aspersa*) – a common land snail

The three most universal features defining modern molluscs are a mantle with a significant cavity used for breathing and excretion, the presence of a radula (except for bivalves), and the structure of the nervous system. Other than these common elements, molluscs express great morphological diversity, so many textbooks base their descriptions on a "hypothetical ancestral mollusc". This has a single, "limpet-like" shell on top, which is made of proteins and chitin reinforced with calcium carbonate, and is secreted by a mantle covering the whole upper surface. The underside of the animal consists of a single muscular "foot". Although molluscs are coelomates, the coelom tends to be small. The main body cavity is a hemocoel through which blood circulates; as such, their circulatory systems are mainly open. The "generalized" mollusc's feeding system consists of a rasping "tongue", the radula, and a complex digestive system in which exuded mucus and microscopic, muscle-powered "hairs" called cilia play various important roles. The generalized mollusc has two paired nerve cords, or three in bivalves. The brain, in species that have one, encircles the esophagus. Most molluscs have eyes, and all have sensors to detect chemicals, vibrations, and touch. The simplest type of molluscan reproductive system relies on external fertilization, but more complex variations occur. All produce eggs, from which may emerge trochophore larvae, more complex veliger larvae, or miniature adults.

Good evidence exists for the appearance of gastropods, cephalopods and bivalves in the Cambrian period, 541 to 485.4 million years ago. However, the evolutionary history both of molluscs' emergence from the ancestral Lophotrochozoa and of their diversification into the well-known living and fossil forms are still subjects of vigorous debate among scientists.

Molluscs have been and still are an important food source for anatomically modern humans. There is a risk of food poisoning from toxins which can accumulate in certain molluscs under specific conditions, however, and because of this, many countries have regulations to reduce this risk. Molluscs have, for centuries, also been the source of important luxury goods, notably pearls, mother of pearl, Tyrian purple dye, and sea silk. Their shells have also been used as money in some preindustrial societies.

Mollusc species can also represent hazards or pests for human activities. The bite of the blue-ringed octopus is often fatal, and that of *Octopus apollyon* causes inflammation that can last for over a month. Stings from a few species of large tropical cone shells can also kill, but their sophisticated, though easily produced, venoms have become important tools in neurological research. Schistosomiasis (also known as bilharzia, bilharziosis or snail fever) is transmitted to humans via water snail hosts, and affects about 200 million people. Snails and slugs can also be serious agricultural pests, and accidental or deliberate introduction of some snail species into new environments has seriously damaged some ecosystems.

Definition

The most universal features of the body structure of molluscs are a mantle with a significant cavity used for breathing and excretion, and the organization of the nervous system. Many have a calcareous shell.

Molluscs have developed such a varied range of body structures, it is difficult to find synapomorphies (defining characteristics) to apply to all modern groups. The most general characteristic of molluscs is they are unsegmented and bilaterally symmetrical. The following are present in all modern molluscs:

- The dorsal part of the body wall is a mantle (or pallium) which secretes calcareous spicules, plates or shells. It overlaps the body with enough spare room to form a mantle cavity.
- The anus and genitals open into the mantle cavity.
- There are two pairs of main nerve cords.

Other characteristics that commonly appear in textbooks have significant exceptions:

	Whether characteristic is found in these classes of Molluscs						
Supposed universal Molluscan characteristic	Aplacophora	Polyplacophora	Monoplacophora	Gastropoda	Cephalopoda	Bivalvia	Scaphopoda
Radula, a rasping "tongue" with chitinous teeth	Absent in 20% of Neomeniomorpha	Yes	Yes	Yes	Yes	No	Internal, cannot extend beyond body
Broad, muscular foot	Reduced or absent	Yes	Yes	Yes	Modified into arms	Yes	Small, only at "front" end
Dorsal concentration of internal organs (visceral mass)	Not obvious	Yes	Yes	Yes	Yes	Yes	Yes
Large digestive ceca	No ceca in some Aplacophora	Yes	Yes	Yes	Yes	Yes	No
Large complex metanephridia ("kidneys")	None	Yes	Yes	Yes	Yes	Yes	Small, simple

One or more valves/ shells	Primitive forms, yes; modern forms, no	Yes	Yes	Snails, yes; slugs, mostly yes (internal vestigial)	Octopuses, no; cuttlefish, nautilus, squid, yes	Yes	Yes
Odontophore	Yes	Yes	Yes	Yes	Yes	No	Yes

Diversity

About 80% of all known mollusc species are gastropods (snails and slugs), including this cowry (a sea snail).

Estimates of accepted described living species of molluscs vary from 50,000 to a maximum of 120,000 species. In 1969 David Nicol estimated the probable total number of living mollusc species at 107,000 of which were about 12,000 fresh-water gastropods and 35,000 terrestrial. The Bivalvia would comprise about 14% of the total and the other five classes less than 2% of the living molluscs. In 2009, Chapman estimated the number of described living species at 85,000. Haszprunar in 2001 estimated about 93,000 named species, which include 23% of all named marine organisms. Molluscs are second only to arthropods in numbers of living animal species—far behind the arthropods' 1,113,000 but well ahead of chordates' 52,000. About 200,000 living species in total are estimated, and 70,000 fossil species, although the total number of mollusc species ever to have existed, whether or not preserved, must be many times greater than the number alive today.

Molluscs have more varied forms than any other animal phylum. They include snails, slugs and other gastropods; clams and other bivalves; squids and other cephalopods; and other lesser-known but similarly distinctive subgroups. The majority of species still live in the oceans, from the seashores to the abyssal zone, but some form a significant part of the freshwater fauna and the terrestrial ecosystems. Molluscs are extremely diverse in tropical and temperate regions, but can be found at all latitudes. About 80% of all known mollusc species are gastropods. Cephalopoda such as squid, cuttlefish, and octopuses are among the neurologically most advanced of all invertebrates. The giant squid, which until recently had not been observed alive in its adult form, is one of the largest invertebrates, but a recently caught specimen of the colossal squid, 10 m (33 ft) long and weighing 500 kg (1,100 lb), may have overtaken it.

Freshwater and terrestrial molluscs appear exceptionally vulnerable to extinction. Estimates of the numbers of nonmarine molluscs vary widely, partly because many regions have not been thoroughly surveyed. There is also a shortage of specialists who can identify all the animals in any one area to species. However, in 2004 the IUCN Red List of Threatened Species included nearly 2,000 endangered nonmarine molluscs. For comparison, the great majority of mollusc species are marine, but only 41 of these appeared on the 2004 Red List. About 42% of recorded extinctions since the year 1500 are of molluscs, consisting almost entirely of nonmarine species.

Hypothetical Ancestral Mollusc

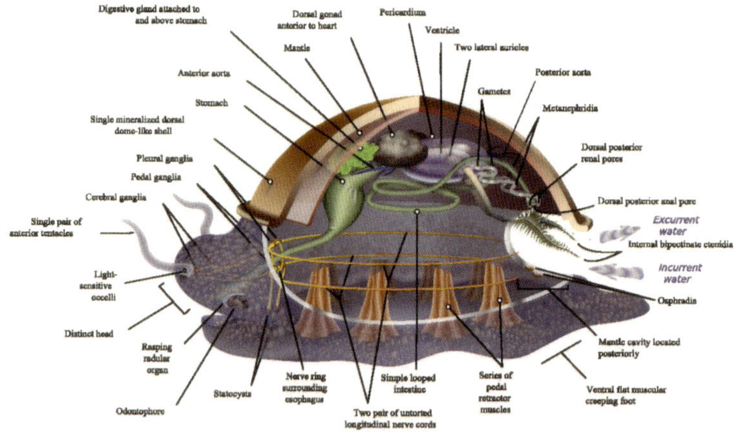

Anatomical diagram of a hypothetical ancestral mollusc

Because of the great range of anatomical diversity among molluscs, many textbooks start the subject of molluscan anatomy by describing what is called an *archi-mollusc, hypothetical generalized mollusc*, or *hypothetical ancestral mollusc* (*HAM*) to illustrate the most common features found within the phylum. The depiction is visually rather similar to modern monoplacophorans.

The generalized mollusc is bilaterally symmetrical and has a single, "limpet-like" shell on top. The shell is secreted by a mantle covering the upper surface. The underside consists of a single muscular "foot". The visceral mass, or visceropallium, is the soft, nonmuscular metabolic region of the mollusc. It contains the body organs.

Mantle and Mantle Cavity

The mantle cavity, a fold in the mantle, encloses a significant amount of space. It is lined with epidermis, and is exposed, according to habitat, to sea, fresh water or air. The cavity was at the rear in the earliest molluscs, but its position now varies from group to group. The anus, a pair of osphradia (chemical sensors) in the incoming "lane", the hindmost pair of gills and the exit openings of the nephridia ("kidneys") and gonads (reproductive organs) are in the mantle cavity. The whole soft body of bivalves lies within an enlarged mantle cavity.

Shell

The mantle edge secretes a shell (secondarily absent in a number of taxonomic groups, such as the nudibranchs) that consists of mainly chitin and conchiolin (a protein hardened with calcium carbonate), except the outermost layer, which in almost all cases is all conchiolin. Molluscs never use phosphate to construct their hard parts, with the questionable exception of *Cobcrephora*. While most mollusc shells are composed mainly of aragonite, those gastropods that lay eggs with a hard shell use calcite (sometimes with traces of aragonite) to construct the eggshells.

The shell consists of three layers: the outer layer (the periostracum) made of organic matter, a middle layer made of columnar calcite, and an inner layer consisting of laminated calcite, often nacreous.

Foot

The underside consists of a muscular foot, which has adapted to different purposes in different classes. The foot carries a pair of statocysts, which act as balance sensors. In gastropods, it secretes mucus as a lubricant to aid movement. In forms having only a top shell, such as limpets, the foot acts as a sucker attaching the animal to a hard surface, and the vertical muscles clamp the shell down over it; in other molluscs, the vertical muscles pull the foot and other exposed soft parts into the shell. In bivalves, the foot is adapted for burrowing into the sediment; in cephalopods it is used for jet propulsion, and the tentacles and arms are derived from the foot.

Circulatory System

Molluscs' circulatory systems are mainly open. Although molluscs are coelomates, their coeloms are reduced to fairly small spaces enclosing the heart and gonads. The main body cavity is a hemocoel through which blood and coelomic fluid circulate and which encloses most of the other internal organs. These hemocoelic spaces act as an efficient hydrostatic skeleton. The blood contains the respiratory pigment hemocyanin as an oxygen-carrier. The heart consists of one or more pairs of atria (auricles), which receive oxygenated blood from the gills and pump it to the ventricle, which pumps it into the aorta (main artery), which is fairly short and opens into the hemocoel.

The atria of the heart also function as part of the excretory system by filtering waste products out of the blood and dumping it into the coelom as urine. A pair of nephridia ("little kidneys") to the rear of and connected to the coelom extracts any re-usable materials from the urine and dumps additional waste products into it, and then ejects it via tubes that discharge into the mantle cavity.

Respiration

Most molluscs have only one pair of gills, or even only a singular gill. Generally, the gills are rather like feathers in shape, although some species have gills with filaments on only one side. They divide the mantle cavity so water enters near the bottom and exits near the top. Their filaments have three kinds of cilia, one of which drives the water current through the mantle cavity, while the other two help to keep the gills clean. If the osphradia detect noxious chemicals or possibly sediment entering the mantle cavity, the gills' cilia may stop beating until the unwelcome intrusions have ceased. Each gill has an incoming blood vessel connected to the hemocoel and an outgoing one to the heart.

Eating, Digestion, and Excretion

Members of the mollusc family use intracellular digestion to function. Most molluscs have muscular mouths with radulae, "tongues", bearing many rows of chitinous teeth, which are replaced from the rear as they wear out. The radula primarily functions to scrape bacteria and algae off rocks, and is associated with the odontophore, a cartilaginous supporting organ. The radula is unique to the molluscs and has no equivalent in any other animal.

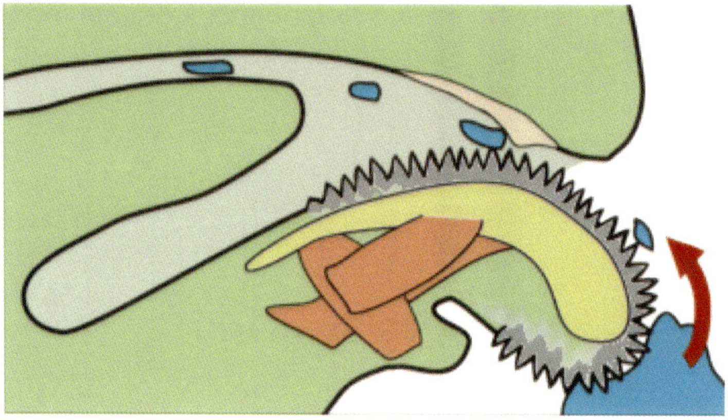

Snail radula at work

= Food = Radula = Muscles = Odontophore "belt"

Molluscs' mouths also contain glands that secrete slimy mucus, to which the food sticks. Beating cilia (tiny "hairs") drive the mucus towards the stomach, so the mucus forms a long string called a "food string".

At the tapered rear end of the stomach and projecting slightly into the hindgut is the prostyle, a backward-pointing cone of feces and mucus, which is rotated by further cilia so it acts as a bobbin, winding the mucus string onto itself. Before the mucus string reaches the prostyle, the acidity of the stomach makes the mucus less sticky and frees particles from it.

The particles are sorted by yet another group of cilia, which send the smaller particles, mainly minerals, to the prostyle so eventually they are excreted, while the larger ones, mainly food, are sent to the stomach's cecum (a pouch with no other exit) to be digested. The sorting process is by no means perfect.

Periodically, circular muscles at the hindgut's entrance pinch off and excrete a piece of the prostyle, preventing the prostyle from growing too large. The anus, in the part of the mantle cavity, is swept by the outgoing "lane" of the current created by the gills. Carnivorous molluscs usually have simpler digestive systems.

As the head has largely disappeared in bivalves, the mouth has been equipped with labial palps (two on each side of the mouth) to collect the detritus from its mucus.

Nervous System

The cephalic molluscs have two pairs of main nerve cords organized around a number of paired ganglia, the visceral cords serving the internal organs and the pedal ones serving the foot. Most pairs of corresponding ganglia on both sides of the body are linked by commissures (relatively large bundles of nerves). The ganglia above the gut are the cerebral, the pleural, and the visceral, which are located above the esophagus (gullet). The pedal ganglia, which control the foot, are below the esophagus and their commissure and connectives to the cerebral and pleural ganglia surround the esophagus in a circumesophageal nerve ring or *nerve collar*.

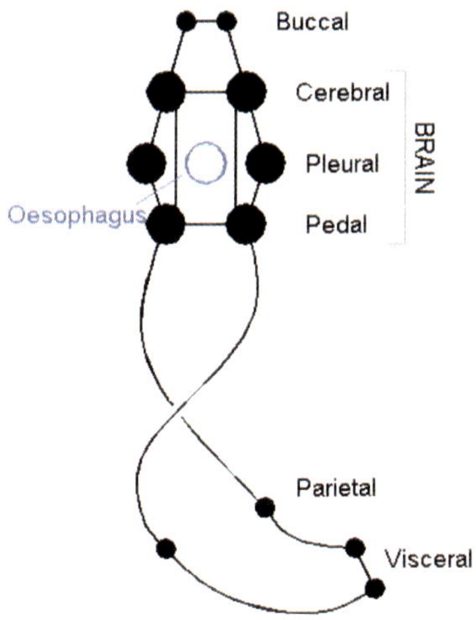

Simplified diagram of the mollusc nervous system

The acephalic molluscs (i.e., bivalves) also have this ring but it is less obvious and less important. The bivalves have only three pairs of ganglia— cerebral, pedal, and visceral— with the visceral as the largest and most important of the three functioning as the principal center of "thinking". Some such as the scallops have eyes around the edges of their shells which connect to a pair of looped nerves and which provide the ability to distinguish between light and shadow.

Reproduction

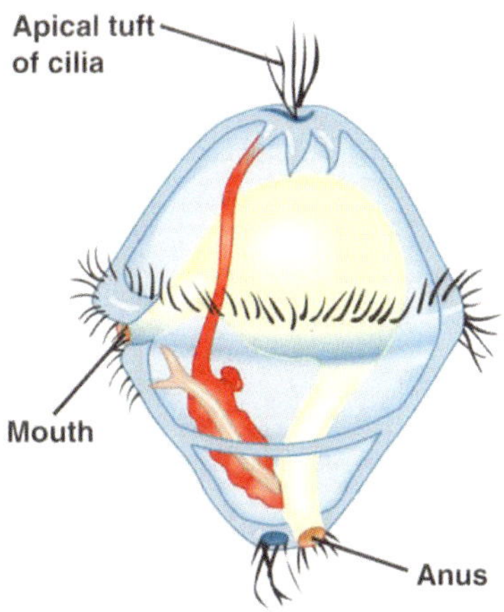

Structure of trochophore larva

Trochophore larva

The simplest molluscan reproductive system relies on external fertilization, but with more complex variations. All produce eggs, from which may emerge trochophore larvae, more complex veliger larvae, or miniature adults. Two gonads sit next to the coelom, a small cavity that surrounds the heart, into which they shed ova or sperm. The nephridia extract the gametes from the coelom and emit them into the mantle cavity. Molluscs that use such a system remain of one sex all their lives and rely on external fertilization. Some molluscs use internal fertilization and/or are hermaphrodites, functioning as both sexes; both of these methods require more complex reproductive systems.

The most basic molluscan larva is a trochophore, which is planktonic and feeds on floating food particles by using the two bands of cilia around its "equator" to sweep food into the mouth, which uses more cilia to drive them into the stomach, which uses further cilia to expel undigested remains through the anus. New tissue grows in the bands of mesoderm in the interior, so the apical tuft and anus are pushed further apart as the animal grows. The trochophore stage is often succeeded by a veliger stage in which the prototroch, the "equatorial" band of cilia nearest the apical tuft, develops into the velum ("veil"), a pair of cilia-bearing lobes with which the larva swims. Eventually, the larva sinks to the seafloor and metamorphoses into the adult form. While metamorphosis is the usual state in molluscs, the cephalopods differ in exhibiting direct development: the hatchling is a 'miniaturized' form of the adult.

Ecology

Feeding

Most molluscs are herbivorous, grazing on algae or filter feeders. For those grazing, two feeding strategies are predominant. Some feed on microscopic, filamentous algae, often using their radula as a 'rake' to comb up filaments from the sea floor. Others feed on macroscopic 'plants' such as kelp, rasping the plant surface with its radula. To employ this strategy, the plant has to be large enough for the mollusc to 'sit' on, so smaller macroscopic plants are not as often eaten as their larger counterparts. Filter feeders are molluscs that feed by straining suspended matter and food particle from water, typically by passing the water over their gills. Most bivalves are filter feeders.

Cephalopods are primarily predatory, and the radula takes a secondary role to the jaws and tentacles in food acquisition. The monoplacophoran *Neopilina* uses its radula in the usual fashion, but its diet includes protists such as the xenophyophore *Stannophyllum*. Sacoglossan sea-slugs suck the sap from algae, using their one-row radula to pierce the cell walls, whereas dorid nudibranchs and some Vetigastropoda feed on sponges and others feed on hydroids.

Classification

Opinions vary about the number of classes of molluscs; for example, the table below shows eight living classes, and two extinct ones. Although they are unlikely to form a clade, some older works combine the Caudofoveata and solenogasters into one class, the Aplacophora. Two of the commonly recognized "classes" are known only from fossils.

Class	Major organisms	Described living species	Distribution
Gastropoda	All the snails and slugs including abalone, limpets, conch, nudibranchs, sea hares, sea butterfly	70,000	marine, freshwater, land
Bivalvia	clams, oysters, scallops, geoducks, mussels	20,000	marine, freshwater
Polyplacophora	chitons	1,000	rocky tidal zone and seabed
Cephalopoda	squid, octopus, cuttlefish, nautilus, spirula	900	marine
Scaphopoda	tusk shells	500	marine 6–7,000 metres (20–22,966 ft)

Solenogastres	worm-like organisms	200	seabed 200–3,000 metres (660–9,840 ft)
Caudofoveata	worm-like organisms	120	seabed 200–3,000 metres (660–9,840 ft)
Monoplacophora	An ancient lineage of molluscs with cap-like shells	31	seabed 1,800–7,000 metres (5,900–23,000 ft); one species 200 metres (660 ft)
Rostroconchia	fossils; probable ancestors of bivalves	extinct	marine
Helcionelloida	fossils; snail-like organisms such as *Latouchella*	extinct	marine

Classification into higher taxa for these groups has been and remains problematic. A phylogenetic study suggests the Polyplacophora form a clade with a monophyletic Aplacophora. Additionally, it suggests a sister taxon relationship exists between the Bivalvia and the Gastropoda. Tentaculita may also be in Mollusca.

Evolution

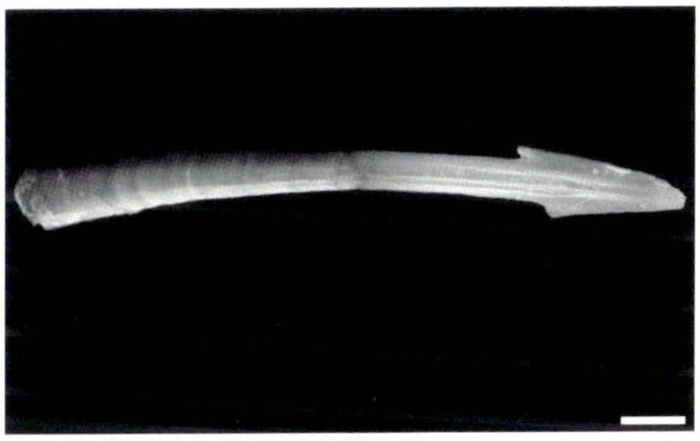

The use of love darts by the land snail *Monachoides vicinus* is a form of sexual selection

Fossil Record

Good evidence exists for the appearance of gastropods (e.g. *Aldanella*), cephalopods (e.g. *Plectronoceras, Nectocaris*) and bivalves (*Pojetaia, Fordilla*) towards the middle of the Cambrian period, c. 500 million years ago, though arguably each of these may belong only to the stem lineage of their respective classes. However, the evolutionary history both of the emergence of molluscs from the ancestral group Lophotrochozoa, and of their diversification into the well-known living and fossil forms, is still vigorously debated.

Debate occurs about whether some Ediacaran and Early Cambrian fossils really are molluscs. *Kimberella*, from about 555 million years ago, has been described by some paleontologists as "mollusc-like", but others are unwilling to go further than "probable bilaterian", if that.

There is an even sharper debate about whether *Wiwaxia*, from about 505 million years ago, was a mollusc, and much of this centers on whether its feeding apparatus was a type of radula or more similar to that of some polychaete worms. Nicholas Butterfield, who opposes the idea that *Wiwaxia* was a mollusc, has written that earlier microfossils from 515 to 510 million years ago are fragments of a genuinely mollusc-like radula. This appears to contradict the concept that the ancestral molluscan radula was mineralized.

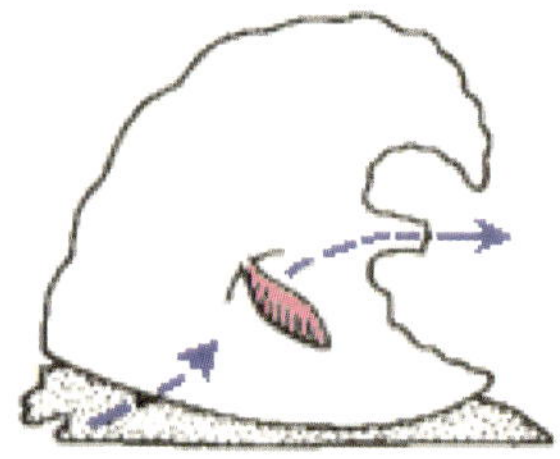

The tiny Helcionellid fossil Yochelcionella is thought to be an early mollusc

Spirally coiled shells appear in many gastropods.

However, the Helcionellids, which first appear over 540 million years ago in Early Cambrian rocks from Siberia and China, are thought to be early molluscs with rather snail-like shells. Shelled molluscs therefore predate the earliest trilobites. Although most helcionellid fossils are only a few millimeters long, specimens a few centimeters long have also been found, most with more limpet-like shapes. The tiny specimens have been suggested to be juveniles and the larger ones adults.

Some analyses of helcionellids concluded these were the earliest gastropods. However, other scientists are not convinced these Early Cambrian fossils show clear signs of the torsion that identifies modern gastropods twists the internal organs so the anus lies above the head.

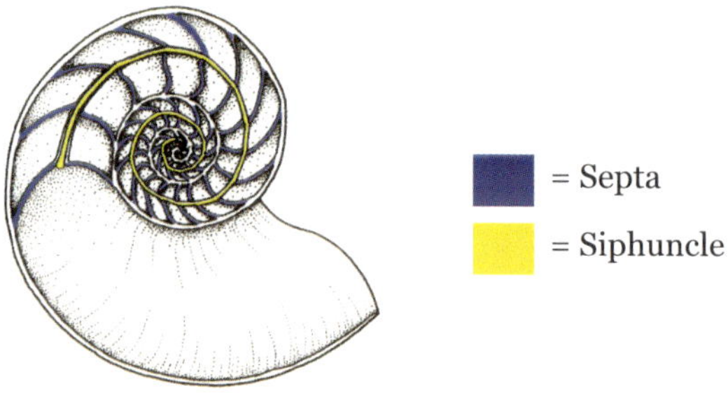

Septa and siphuncle in nautiloid shell

Volborthella, some fossils of which predate 530 million years ago, was long thought to be a cephalopod, but discoveries of more detailed fossils showed its shell was not secreted, but built from grains of the mineral silicon dioxide (silica), and it was not divided into a series of compartments by septa as those of fossil shelled cephalopods and the living *Nautilus* are. *Volborthella*'s classification is uncertain. The Late Cambrian fossil *Plectronoceras* is now thought to be the earliest clearly cephalopod fossil, as its shell had septa and a siphuncle, a strand of tissue that *Nautilus* uses to remove water from compartments it has vacated as it grows, and which is also visible in fossil ammonite shells. However, *Plectronoceras* and other early cephalopods crept along the seafloor instead of swimming, as their shells contained a "ballast" of stony deposits on what is thought to be the underside, and had stripes and blotches on what is thought to be the upper surface. All cephalopods with external shells except the nautiloids became extinct by the end of the Cretaceous period 65 million years ago. However, the shell-less Coleoidea (squid, octopus, cuttlefish) are abundant today.

The Early Cambrian fossils *Fordilla* and *Pojetaia* are regarded as bivalves. "Modern-looking" bivalves appeared in the Ordovician period, 488 to 443 million years ago. One bivalve group, the rudists, became major reef-builders in the Cretaceous, but became extinct in the Cretaceous–Paleogene extinction event. Even so, bivalves remain abundant and diverse.

The Hyolitha are a class of extinct animals with a shell and operculum that may be molluscs. Authors who suggest they deserve their own phylum do not comment on the position of this phylum in the tree of life.

Phylogeny

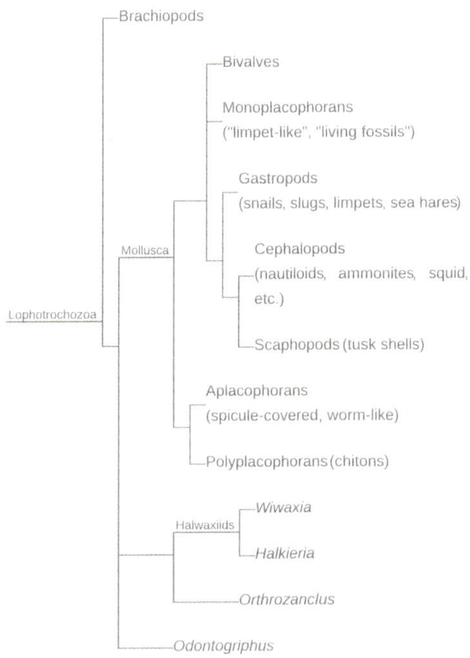

A possible "family tree" of molluscs. Does not include annelid worms as the analysis concentrated on fossilizable "hard" features.

The phylogeny (evolutionary "family tree") of molluscs is a controversial subject. In addition to the debates about whether *Kimberella* and any of the "halwaxiids" were molluscs or closely related to molluscs, debates arise about the relationships between the classes of living molluscs. In fact, some groups traditionally classified as molluscs may have to be redefined as distinct but related.

Molluscs are generally regarded members of the Lophotrochozoa, a group defined by having trochophore larvae and, in the case of living Lophophorata, a feeding structure called a lophophore. The other members of the Lophotrochozoa are the annelid worms and seven marine phyla. The diagram on the right summarizes a phylogeny presented in 2007.

Because the relationships between the members of the family tree are uncertain, it is difficult to identify the features inherited from the last common ancestor of all molluscs. For example, it is uncertain whether the ancestral mollusc was metameric (composed of repeating units)—if it was, that would suggest an origin from an annelid-like worm. Scientists disagree about this: Giribet and colleagues concluded, in 2006, the repetition of gills and of the foot's retractor muscles were later developments, while in 2007, Sigwart concluded the ancestral mollusc was metameric, and it had a foot used for creeping and a "shell" that was mineralized. In one particular branch of the family tree, the shell of conchiferans is thought to have evolved from the spicules (small spines) of aplacophorans; but this is difficult to reconcile with the embryological origins of spicules.

The molluscan shell appears to have originated from a mucus coating, which eventually stiffened into a cuticle. This would have been impermeable and thus forced the development of more sophisticated respiratory apparatus in the form of gills. Eventually, the cuticle would have become mineralized, using the same genetic machinery (engrailed) as most other bilaterian skeletons. The first mollusc shell almost certainly was reinforced with the mineral aragonite.

The evolutionary relationships within the molluscs are also debated, and the diagrams below show two widely supported reconstructions:

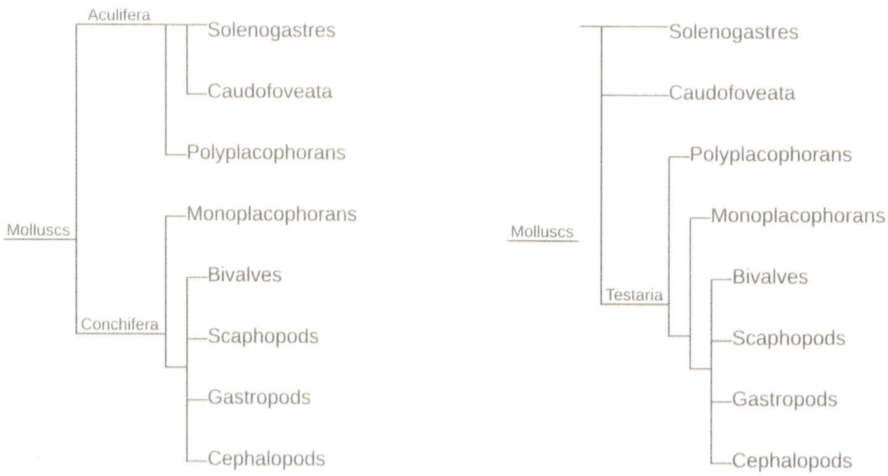

The "Aculifera" hypothesis The "Testaria" hypothesis

Morphological analyses tend to recover a conchiferan clade that receives less support from molecular analyses, although these results also lead to unexpected paraphylies, for instance scattering the bivalves throughout all other mollusc groups.

However, an analysis in 2009 using both morphological and molecular phylogenetics comparisons concluded the molluscs are not monophyletic; in particular, Scaphopoda and Bivalvia are both separate, monophyletic lineages unrelated to the remaining molluscan classes; the traditional phylum Mollusca is polyphyletic, and it can only be made monophyletic if scaphopods and bivalves are excluded. A 2010 analysis recovered the traditional conchiferan and aculiferan groups, and showed molluscs were monophyletic, demonstrating that available data for solenogastres was contaminated. Current molecular data are insufficient to constrain the molluscan phylogeny, and since the methods used to determine the confidence in clades are prone to overestimation, it is risky to place too much emphasis even on the areas of which different studies agree. Rather than eliminating unlikely relationships, the latest studies add new permutations of internal molluscan relationships, even bringing the conchiferan hypothesis into question.

Human Interaction

For millennia, molluscs have been a source of food for humans, as well as important luxury goods, notably pearls, mother of pearl, Tyrian purple dye, sea silk, and chemical compounds. Their shells have also been used as a form of currency in some preindustrial societies. A number of species of molluscs can bite or sting humans, and some have become agricultural pests.

Uses by Humans

Molluscs, especially bivalves such as clams and mussels, have been an important food source since at least the advent of anatomically modern humans, and this has often resulted in overfishing. Other commonly eaten molluscs include octopuses and squids, whelks, oysters, and scallops. In 2005, China accounted for 80% of the global mollusc catch, netting almost 11,000,000 tonnes (11,000,000 long tons; 12,000,000 short tons). Within Europe, France remained the industry leader. Some countries regulate importation and handling of molluscs and other seafood, mainly to minimize the poison risk from toxins that can sometimes accumulate in the animals.

Saltwater pearl oyster farm in Seram, Indonesia

Most molluscs with shells can produce pearls, but only the pearls of bivalves and some gastropods, whose shells are lined with nacre, are valuable. The best natural pearls are produced by marine pearl oysters, *Pinctada margaritifera* and *Pinctada mertensi*, which live in the tropical and subtropical waters of the Pacific Ocean. Natural pearls form when a small foreign object gets stuck between the mantle and shell.

The two methods of culturing pearls insert either "seeds" or beads into oysters. The "seed" method uses grains of ground shell from freshwater mussels, and overharvesting for this purpose has endangered several freshwater mussel species in the southeastern United States. The pearl industry is so important in some areas, significant sums of money are spent on monitoring the health of farmed molluscs.

Other luxury and high-status products were made from molluscs. Tyrian purple, made from the ink glands of murex shells, "... fetched its weight in silver" in the fourth century BC, according to Theopompus. The discovery of large numbers of *Murex* shells on Crete suggests the Minoans may have pioneered the extraction of "imperial purple" during the Middle Minoan period in the 20th–18th centuries BC, centuries before the Tyrians. Sea silk is a fine, rare, and valuable fabric produced from the long silky threads (byssus) secreted by several bivalve molluscs, particularly *Pinna nobilis*, to attach themselves to the sea bed. Procopius, writing on the Persian wars *circa* 550 CE, "stated that the five hereditary satraps (governors) of Armenia who received their insignia from the Roman Emperor were given chlamys (or cloaks) made from *lana pinna*. Apparently, only the ruling classes were allowed to wear these chlamys."

Byzantine Emperor Justinian I clad in Tyrian purple and wearing numerous pearls

Mollusc shells, including those of cowries, were used as a kind of money (shell money) in several preindustrial societies. However, these "currencies" generally differed in important ways from the standardized government-backed and -controlled money familiar to industrial societies. Some shell "currencies" were not used for commercial transactions, but mainly as social status displays at important occasions, such as weddings. When used for commercial transactions, they functioned as commodity money, as a tradable commodity whose value differed from place to place, often as a result of difficulties in transport, and which was vulnerable to incurable inflation if more efficient transport or "goldrush" behavior appeared.

Bioindicators

Bivalve molluscs are used as bioindicators to monitor the health of aquatic environments in both fresh water and the marine environments. Their population status or structure, physiology, behaviour or the level of contamination with elements or compounds can indicate the state of contamination status of the ecosystem. They are particularly useful since they are sessile so that they are representative of the environment where they are sampled or placed. Potamopyrgus antipodarum is used by some water treatment plants to test for estrogen-mimicking pollutants from industrial agriculture.

Harmful to Humans

Stings and Bites

The blue-ringed octopus's rings are a warning signal; this octopus is alarmed, and its bite can kill.

Some molluscs sting or bite, but deaths from mollusc venoms total less than 10% of those from jellyfish stings.

All octopuses are venomous, but only a few species pose a significant threat to humans. Blue-ringed octopuses in the genus *Hapalochlaena*, which live around Australia and New Guinea, bite humans only if severely provoked, but their venom kills 25% of human victims. Another tropical species, *Octopus apollyon*, causes severe inflammation that can last for over a month even if treated correctly, and the bite of *Octopus rubescens* can cause necrosis that lasts longer than one month if untreated, and headaches and weakness persisting for up to a week even if treated.

Live cone snails can be dangerous to shell collectors, but are useful to neurology researchers.

All species of cone snails are venomous and can sting painfully when handled, although many species are too small to pose much of a risk to humans, and only a few fatalities have been reliably reported. Their venom is a complex mixture of toxins, some fast-acting and others slower but deadlier. The effects of individual cone-shell toxins on victims' nervous systems are so precise as to be useful tools for research in neurology, and the small size of their molecules makes it easy to synthesize them.

Disease Vectors

Schistosomiasis (also known as bilharzia, bilharziosis or snail fever), a disease caused by the fluke worm *Schistosoma*, is "second only to malaria as the most devastating parasitic disease in tropical countries. An estimated 200 million people in 74 countries are infected with the disease – 100 million in Africa alone." The parasite has 13 known species, two of which infect humans. The parasite itself is not a mollusc, but all the species have freshwater snails as intermediate hosts.

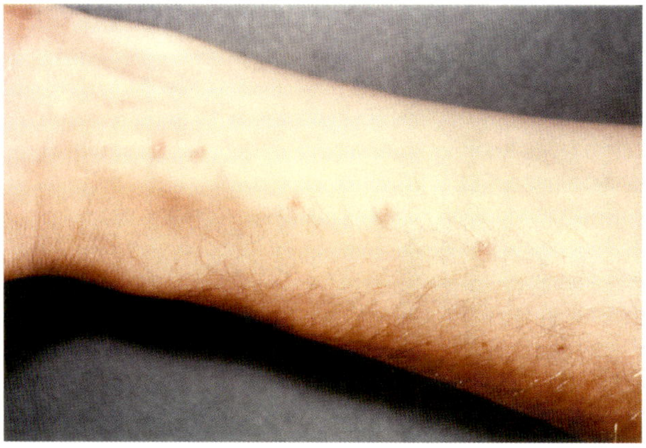

Skin vesicles created by the penetration of *Schistosoma*.

Pests

Some species of molluscs, particularly certain snails and slugs, can be serious crop pests, and when introduced into new environments, can unbalance local ecosystems. One such pest, the giant African snail *Achatina fulica*, has been introduced to many parts of Asia, as well as to many islands in the Indian Ocean and Pacific Ocean. In the 1990s, this species reached the West Indies. Attempts to control it by introducing the predatory snail *Euglandina rosea* proved disastrous, as the predator ignored *Achatina fulica* and went on to extirpate several native snail species, instead.

Subdivisions of Malacology

Conchology

Conchology is the study of mollusc shells. Conchology is one aspect of malacology, the study of molluscs; however, malacology is the study of molluscs as whole organisms, whereas conchology is confined to the study of their shells. It includes the study of land and freshwater mollusc shells as well as seashells and extends to the study of a gastropod's operculum.

Calliostoma tigris

Shell of *Lobatus gigas*, the queen conch

Conchology is now sometimes seen as an archaic study, because relying on only one aspect of an organism's morphology can be misleading. However, a shell often gives at least some insight into molluscan taxonomy, and historically the shell was often the only part of exotic species that was available for study. Even in current museum collections it is common for the dry material (shells) to greatly exceed the amount of material that is preserved whole in alcohol.

Conchologists mainly deal with four molluscan orders: the gastropods (snails), bivalves (clams), Polyplacophora (chitons) and Scaphopoda (tusk shells). Cephalopods only have small internal shells, with the exception of the Nautiloidea. Some groups, such as the sea slug nudibranchs, have lost their shells altogether, while in others it has been replaced by a protein support structure.

Shell Collecting versus Conchology

A vendor in Tanzania with a variety of large seashells for sale

The terms *shell collector* and *conchologist* can be regarded as two distinct categories. Not all shell collectors are conchologists; some are primarily concerned with the aesthetic value of shells instead of their scientific study. It is also true that not all conchologists are shell collectors; this type of research only requires access to private or institutional shell collections. There is some debate in the conchological community, with some people regarding all shell collectors (regardless of motivation) as conchologists.

Fake Shells

Shell collectors who purchase shells from dealers may sometimes encounter shells which have been altered to represent new species or rare color varieties. It is claimed that in previous centuries, fake examples of *Epitonium scalare* were created out of rice paste.

Depictions of Shells on Stamps and Coins

Shells have been featured on over 5,000 postage stamps worldwide.

Shells have also been featured on many coins, including the Bahamian dollar (1974), the Cuban peso (1981), the Haitian gourde (1973), the Nepalese rupee (1989) and Philippine peso (1993).

Limacology

Limacology is the branch of zoology which deals with slugs, i.e. shell-less gastropod mollusks. A person that studies limacology is referred to as a limacologist.

However, slugs are an extremely polyphyletic group, thus "limacology" is not a taxonomically accurate term, and it is now rarely used.

Slug

Slug, or land slug, is a common name for any apparently shell-less terrestrial gastropod mollusc. The word *slug* is also often used as part of the common name of any gastropod mollusc that has no shell, a very reduced shell, or only a small internal shell, particularly sea slugs and semislugs (this is in contrast to the common name *snail*, which applies to gastropods that have a coiled shell large enough that the animal can fully retract its soft parts into the shell).

Various taxonomic families of land slugs form part of several quite different evolutionary lineages, which also include snails. Thus, the various families of slugs are not closely related, despite a superficial similarity in the overall body form. The shell-less condition has arisen many times independently during the evolutionary past, and thus the category "slug" is a polyphyletic one.

Description

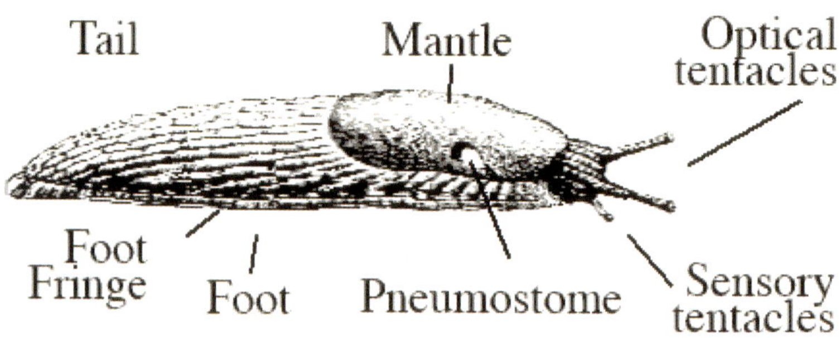

The external anatomy of a slug

The external anatomy of a slug includes the following:

Tentacles: Like other pulmonate land gastropods, the majority of land slugs have two pairs of 'feelers' or *tentacles* on their head. The upper pair is light sensing and has eyespots at the ends, while the lower pair provides the sense of smell. Both pairs are retractable.

Mantle: On top of the slug, behind the head, is the saddle-shaped *mantle,* and under this are the genital opening and anus. On one side (almost always the right hand side) of the mantle is a respiratory opening, which is easy to see when open, but difficult to see when closed. This opening is known as the pneumostome.

Tail: The part of a slug behind the mantle is called the 'tail'.

Keel: Some species of slugs, for example *Tandonia budapestensis*, have a prominent ridge running over their back along the middle of the tail (sometimes along the whole tail, sometimes only the final part). This ridge is called a 'keel'.

Foot: The bottom side of a slug, which is flat, is called the 'foot'. Like almost all gastropods, a slug moves by rhythmic waves of muscular contraction on the underside of its foot. It simultaneously secretes a layer of mucus that it travels on, which helps prevent damage to the foot tissues. Around the edge of the foot in some taxa is a structure called the 'foot fringe'.

Vestigial shell: Most slugs retain a remnant of their shell, which is usually internalized. This organ generally serves as storage for calcium salts, often in conjunction with the digestive glands. An internal shell is present in the Limacidae and Parmacellidae. Adult Philomycidae, Onchidiidae and Veronicellidae lack shells.

Physiology

An active *Ambigolimax* slug in Fremont, California

Slugs' bodies are made up mostly of water and, without a full-sized shell, their soft tissues are prone to desiccation. They must generate protective mucus to survive. Many species are most active just after rain because of the moist ground. In drier conditions, they hide in damp places such as under tree bark, fallen logs, rocks and man-made structures, such as planters, to help retain body moisture. Like all other gastropods, they undergo torsion (a 180° twisting of the internal organs) during development. Internally, slug anatomy clearly shows the effects of this rotation—but externally, the bodies of slugs appear more or less symmetrical, except for the positioning of the pneumostome, which is on one side of the animal, normally the right-hand side.

Slugs produce two types of mucus: one is thin and watery, and the other thick and sticky. Both kinds are hygroscopic. The thin mucus spreads from the foot's centre to its edges, whereas the thick mucus spreads from front to back. Slugs also produce thick mucus that coats the whole body of the animal. The mucus secreted by the foot contains fibres that help prevent the slug from slipping down vertical surfaces. The "slime trail" a slug leaves behind has some secondary effects: other slugs coming across a slime trail can recognise the slime trail as produced by one of the same species, which is useful in finding a mate. Following a slime trail is also part of the hunting behaviour of some carnivorous slugs. Body mucus provides some protection against predators, as it can make the slug hard to pick up and hold by a bird's beak, for example, and the mucus itself can be distasteful. Some species of slug, such as *Limax maximus*, secrete slime cords to suspend a pair during copulation.

Reproduction

Deroceras juvenile with eggs of another (larger) slug species, most likely Arionidae

Slugs are hermaphrodites, having both female and male reproductive organs. Once a slug has located a mate, they encircle each other and sperm is exchanged through their protruded genitalia. A few days later, the slugs lay approximately thirty eggs in a hole in the ground, or beneath the cover of an object such as a fallen log.

Apophallation has been reported only in some species of banana slug (*Ariolimax*) and one species of *Deroceras*. In the banana slugs, the penis is trapped inside the body of the partner. Apophallation allows the slugs to separate themselves by one or both of the slugs chewing off the other's or their own penis. Once the penis has been lost, banana slugs are still able to mate using only the female parts of the reproductive system.

Ecology

Slugs play an important role in the ecosystem by eating decaying plant material and fungi. Most carnivorous slugs on occasion also eat dead specimens of their own kind.

Feeding Habits

Most species of slugs are generalists, feeding on a broad spectrum of organic materials, including leaves from living plants, lichens, mushrooms, and even carrion. Some slugs are predators and eat other slugs and snails, or earthworms.

Large Slug near Manali, India

Lehmannia feeding on a small fruit in Mexico City

Slugs can feed on a wide variety of vegetables and herbs, including flowers such as petunias, chrysanthemums, daisies, lobelia, lilies, daffodils, narcissus, gentians, primroses, tuberous begonias, hollyhocks, irises, and fruits such as strawberries. They also feed on carrots, peas, apples, and cabbage that are offered as a sole food source.

Slugs from different families are fungivores. It is the case in the Philomycidae (e. g. *Philomycus carolinianus* and *Phylomicus flexuolaris*) and *Ariolimacidae* (*Ariolimax californianus*), which respectively feed on slime molds (myxomycetes) and mushrooms (basidiomycetes). Species of mushroom producing fungi used as food source by slugs include milk-caps, *Lactarius* spp., the oyster mushroom, *Pleurotus ostreatus* and the penny bun, *Boletus edulis*. Other species pertaining to different genera, such as *Agaricus*, *Pleurocybella* and *Russula*, are also eaten by slugs. Slime molds used as food source by slugs include *Stemonitis axifera* and *Symphytocarpus flaccidus*. Some slugs are selective towards certain parts or developmental stages of the fungi they eat, though this is very variable. Depending on the species and other factors, slugs eat only fungi at specific stages of development. Moreover, in other cases, whole mushrooms can be eaten, without any selection or bias towards ontogenetic stages.

Predators

Slugs are preyed upon by myriad vertebrates and invertebrates. The predation of slugs has been the subject of studies for at least a century. Because some species of slugs are considered agricultural pests, research investments have been made to comprehend and investigate potential predators. This is a necessary knowledge to establish biological control strategies.

Vertebrates

Slugs are preyed upon by virtually every major vertebrate group. With many examples among reptiles, birds, mammals, amphibians and fish, vertebrates can occasionally feed on, or be specialised predators of slugs. Fish that feed on slugs include the brown trout, (*Salmo trutta*), which occasionally feeds on *Arion circumscriptus*, an *arionid* slug. Similarly, the shortjaw kokopu (*Galaxias postvectis*) includes slugs in its diet. Amphibians such as frogs and toads have long been regarded as important predators of slugs. Among them are species in the genus *Bufo* (e. g. *Bufo marinus*) and *Ceratophrys*.

Reptiles that feed on slugs include mainly snakes and lizards. Some colubrid snakes are known predators of slugs. Coastal populations of the garter snake,*Thamnophis elegans*, have a specialised diet consisting of slugs, such as *Ariolimax*, while inland populations have a generalized diet. One of its congeners, the Northwestern garter snake (*Thamnophis ordinoides*), is not a specialized predator of slugs but occasionally feed on them. The redbelly snake (*Storeria occipitomaculata*) and the brown snake (*Storeria dekayi*) feed mainly but not solely on slugs, while some species in the genus *Dipsas*, *Sibynomorphus* (e.g. *Sibynomorphus neuwiedi*) and the common slug eater snake (*Duberria lutrix*), are exclusively slug eaters. Several lizards include slugs in their diet. This is the case in the slowworm (*Anguis fragilis*), the bobtail lizard (*Tiliqua rugosa*), the she-oak skink (*Cyclodomorphus casuarinae*) and the common lizard (*Zootoca vivipara*).

Birds that prey upon slugs include common blackbirds (*Turdus merula*), starlings (*Sturnus vulgaris*), rooks (*Corvus frugilegus*), jackdaws (*Corvus monedula*), owls, vultures and ducks. Studies on slug predation also cite fieldfares (feeding on *Deroceras reticulatum*), redwings (feeding on *Limax* and *Arion*), thrushes (on *Limax* and *Arion ater*), red grouse (on *Deroceras* and *Arion hortensis*), game birds, wrynecks (on *Limax flavus*), rock doves and charadriiform birds as slug predators.

Mammals that eat slugs include foxes, badgers and hedgehogs.

Invertebrates

Beetles in the family Carabidae, such as *Carabus violaceus* and *Pterostichus melanarius*, are known to feed on slugs.

Parasites and Parasitoids

Slugs are parasitised by several organisms, including acari and a wide variety of nematodes. The slug mite, *Riccardoella limacum*, is known to parasitise several dozen species of mollusks, including many slugs, such as *Agriolimax agrestis*, *Arianta arbustrum*, *Arion ater*, *Arion hortensis*, *Limax maximus*, *Milax budapestensis*, *Milax gagates*, and *Milax sowerbyi*. *R. limacum* can often be seen swarming about their host's body, and live in its respiratory cavity.

Several species of nematodes are known to parasitise slugs. The nematode worms *Agfa flexilis* and *Angiostoma limacis* respectively live in the salivary glands and rectum of *Limax maximus*. Species of widely known medical importance pertaining to the genus *Angiostrongylus* are also parasites of slugs. Both *Angiostrongylus costaricensis* and *Angiostrongylus cantonensis*, a meningitis-causing nematode, have larval stages that can only live in mollusks, including slugs, such as *Limax maximus*.

Insects such as dipterans are known parasitoids of mollusks. To complete their development, many dipterans use slugs as hosts during their ontogeny. Some species of blow-flies (Calliphoridae) in the genus *Melinda* are known parasitoids of Arionidae, Limacidae and Philomycidae. Flies in the family Phoridae, specially those in the genus *Megaselia*, are parasitoids of Agriolimacidae, including many species of *Deroceras*. House flies in the family Muscidae, mainly those in the genus *Sarcophaga*, are facultative parasitoids of Arionidae.

Behavior

The alarm response posture of the Kerry slug, which is found only in this species

When attacked, slugs can contract their body, making themselves harder and more compact and more still and round. By doing this, they become firmly attached to the substrate. This, combined with the slippery mucus they produce, makes slugs more difficult for predators to grasp. The unpleasant taste of the mucus is also a deterrent. Some species present different response behaviors when attacked, such as the Kerry slug. In contrast to the general behavioral pattern, the Kerry slug retracts its head, lets go of the substrate, rolls up completely, and stays contracted in a ball-like shape. This is a unique feature among all the Arionidae, and among most other slugs. Some slugs can self-amputate (autotomy) a portion of their tail to help the slug escape from a predator. Some slug species hibernate underground during the winter in temperate climates, but in other species, the adults die in the autumn.

Intra- and inter-specific agonistic behavior is documented, but varies greatly among slug species. Slugs often resort to aggression, attacking both conspecifics and individuals from other species when competing for resources. This aggressiveness is also influenced by seasonality, because the availability of resources such as shelter and food may be compromised due to climatic conditions. Slugs are prone to attack during the summer, when the availability of resources is reduced. During winter, the aggressive responses are substituted by a gregarious behavior.

Teuthology

Teuthology is the study of cephalopods. It is a branch of malacology, the study of molluscs, in marine zoology. Those that study teuthology are known as teuthologists.

Cephalopod

A cephalopod is any member of the molluscan class Cephalopoda (Greek plural "head-feet") such as a squid, octopus or nautilus. These exclusively marine animals are characterized by bilateral body symmetry, a prominent head, and a set of arms or tentacles (muscular hydrostats) modified from the primitive molluscan foot. Fishermen sometimes call them inkfish, referring to their common ability to squirt ink. The study of cephalopods is a branch of malacology known as teuthology.

Cephalopods became dominant during the Ordovician period, represented by primitive nautiloids. The class now contains two, only distantly related, extant subclasses: Coleoidea, which includes octopuses, squid, and cuttlefish; and Nautiloidea, represented by *Nautilus* and *Allonautilus*. In the Coleoidea, the molluscan shell has been internalized or is absent, whereas in the Nautiloidea, the external shell remains. About 800 living species of cephalopods have been identified. Two important extinct taxa are the Ammonoidea (ammonites) and Belemnoidea (belemnites).

Distribution

Left: A pair of *Sepia officinalis* in shallow water
Right: *Benthoctopus* sp. on the Davidson Seamount at 2,422 m depth

There are over 800 extant species of cephalopod, although new species continue to be described. An estimated 11,000 extinct taxa have been described, although the soft-bodied nature of cephalopods means they are not easily fossilised.

Cephalopods are found in all the oceans of Earth. None of them can tolerate freshwater, but the brief squid, *Lolliguncula brevis*, found in Chesapeake Bay, is a notable partial exception in that it tolerates brackish water. Cephalopods are thought to be unable to live in freshwater due to multiple biochemical constraints, and in their +400 million year existence have never ventured into fully freshwater habitats.

Cephalopods occupy most of the depth of the ocean, from the abyssal plain to the sea surface. Their diversity is greatest near the equator (~40 species retrieved in nets at 11°N by a diversity study) and decreases towards the poles (~5 species captured at 60°N).

Biology

Nervous System and Behavior

Cephalopods are widely regarded as the most intelligent of the invertebrates, and have well developed senses and large brains (larger than those of gastropods). The nervous system of cephalopods is the most complex of the invertebrates and their brain-to-body-mass ratio falls between that of endothermic and ectothermic vertebrates. Captive cephalopods have also been known to climb out of their aquaria, maneuver a distance of the lab floor, enter another aquarium to feed on the crabs, and return to their own aquarium.

Left: An octopus opening a container with a screw cap Right: Hawaiian bobtail squid, *Euprymna scolopes*, burying itself in the sand, leaving only the eyes exposed

The brain is protected in a cartilaginous cranium. The giant nerve fibers of the cephalopod mantle have been widely used for many years as experimental material in neurophysiology; their large diameter (due to lack of myelination) makes them relatively easy to study compared with other animals.

Many cephalopods are social creatures; when isolated from their own kind, some species have been observed shoaling with fish.

Some cephalopods are able to fly through the air for distances of up to 50 m. While cephalopods are not particularly aerodynamic, they achieve these impressive ranges by jet-propulsion; water continues to be expelled from the funnel while the organism is in the air. The animals spread their fins and tentacles to form wings and actively control lift force with body posture. One species, *Todarodes pacificus*, has been observed spreading tentacles in a flat fan shape with a mucus film between the individual tentacles while another, *Sepioteuthis sepioidea*, has been observed putting the tentacles in a circular arrangement.

Senses

Cephalopods have advanced vision, can detect gravity with statocysts, and have a variety of chemical sense organs. Octopuses use their arms to explore their environment and can use them for depth perception.

Vision

Most cephalopods rely on vision to detect predators and prey, and to communicate with one another. Consequently, cephalopod vision is acute: training experiments have shown that the common octopus can distinguish the brightness, size, shape, and horizontal or vertical orientation of objects. The morphological construction gives cephalopod eyes the same performance as sharks; however, their construction differs, as cephalopods lack a cornea, and have an everted retina. Cephalopods' eyes are also sensitive to the plane of polarization of light. Unlike many other cephalopods, nautiluses do not have good vision; their eye structure is highly developed, but lacks a solid lens. They have a simple "pinhole" eye through which water can pass. Instead of vision, the animal is thought to use olfaction as the primary sense for foraging, as well as locating or identifying potential mates.

The primitive nautilus eye functions similarly to a pinhole camera.

Surprisingly, given their ability to change color, all octopodes and most cephalopods are considered to be color blind. Coleoid cephalopods (octopus, squid, cuttlefish) have a single photoreceptor type and lack the ability to determine color by comparing detected photon intensity across multiple spectral channels. When camouflaging themselves, they use their chromatophores to change brightness and pattern according to the background they see, but their ability to match the specific color of a background may come from cells such as iridophores and leucophores that reflect light from the environment. They also produce visual pigments throughout their body, and may sense light levels directly from their body. Evidence of color vision has been found in the sparkling enope squid (*Watasenia scintillans*), which achieves color vision by the use of three distinct retinal molecules (A1, sensitive to red; A2, to purple, and A4, to yellow?) which bind to its opsin.

A cuttlefish with W-shaped pupils which may help them discriminate colors.

In 2015, a novel mechanism for spectral discrimination in cephalopods was described. This relies on the exploitation of chromatic aberration (wavelength-dependence of focal length). Numerical modeling shows that chromatic aberration can yield useful chromatic information through the dependence of image acuity on accommodation. The unusual off-axis slit and annular pupil shapes in cephalopods enhance this ability.

Photoreception

In 2015, molecular evidence was published indicating that cephalopod chromatophores are photosensitive; reverse transcription polymerase chain reactions (RT-PCR) revealed transcripts encoding rhodopsin and retinochrome within the retinas and skin of the longfin inshore squid (*Doryteuthis pealeii*), and the common cuttlefish (*Sepia officinalis*) and broadclub cuttlefish (*Sepia latimanus*). The authors claim this is the first evidence that cephalopod dermal tissues may possess the required combination of molecules to respond to light.

Hearing

Some squids have been shown to detect sound using their statocysts.

Use of Light

This broadclub cuttlefish (*Sepia latimanus*) can change from camouflage tans and browns (top) to yellow with dark highlights (bottom) in less than a second.

Most cephalopods possess an assemblage of skin components that interact with light. These may include iridophores, leucophores, chromatophores and (in some species) photophores. Chromatophores are colored pigment cells that expand and contract in accordance to produce color and pattern which they can use in a startling array of fashions. As well as providing camouflage with their background, some cephalopods bioluminesce, shining light downwards to disguise their shadows from any predators that may lurk below. The bioluminescence is produced by bacterial symbionts; the host cephalopod is able to detect the light produced by these organisms. Bioluminescence may also be used to entice prey, and some species use colorful displays to impress mates, startle predators, or even communicate with one another.

Coloration

Cephalopods can change their colors and patterns in milliseconds, whether for signalling (both within the species and for warning) or active camouflage, as their chromatophores are expanded or contracted. Although color changes appear to rely primarily on vision input, there is evidence that skin cells, specifically chromatophores, can detect light and adjust to light conditions independently of the eyes. Coloration is typically stronger in near-shore species than those living in the open ocean, whose functions tend to be restricted to disruptive camouflage. Most octopuses mimic select structures in their field of view rather than becoming a composite color of their full background.

Evidence of original coloration has been detected in cephalopod fossils dating as far back as the Silurian; these orthoconic individuals bore concentric stripes, which are thought to have served as camouflage. Devonian cephalopods bear more complex color patterns, of unknown function.

Ink

With the exception of the Nautilidae and the species of octopus belonging to the suborder Cirrina, all known cephalopods have an ink sac, which can be used to expel a cloud of dark ink to confuse predators. This sac is a muscular bag which originated as an extension of the hind gut. It lies beneath the gut and opens into the anus, into which its contents – almost pure melanin – can be squirted; its proximity to the base of the funnel means the ink can be distributed by ejected water as the cephalopod uses its jet propulsion. The ejected cloud of melanin is usually mixed, upon expulsion, with mucus, produced elsewhere in the mantle, and therefore forms a thick cloud, resulting in visual (and possibly chemosensory) impairment of the predator, like a smokescreen. However, a more sophisticated behaviour has been observed, in which the cephalopod releases a cloud, with a greater mucus content, that approximately resembles the cephalopod that released it (this decoy is referred to as a Pseudomorph). This strategy often results in the predator attacking the pseudomorph, rather than its rapidly departing prey.

The ink sac of cephalopods has led to a common name of "inkfish", formerly the pen-and-ink fish.

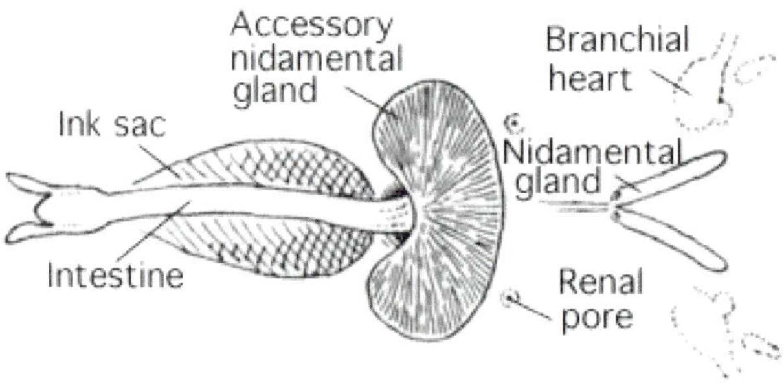

Viscera of *Chtenopteryx sicula*

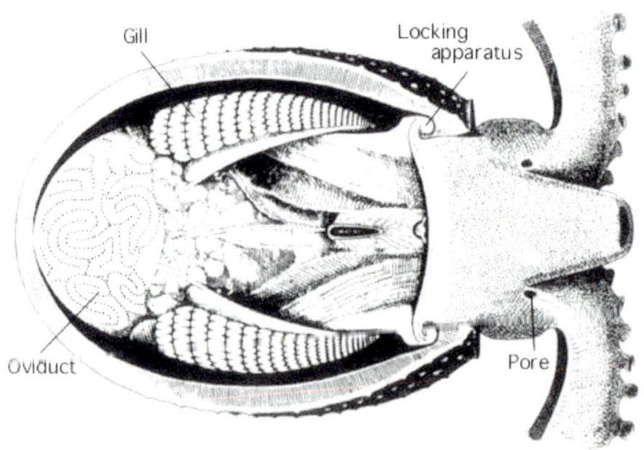

Viscera of *Ocythoe tuberculata*

Circulatory System

Cephalopods are the only mollusks with a closed circulatory system. Coleoids have two gill hearts (also known as branchial hearts) that move blood through the capillaries of the gills. A single systemic heart then pumps the oxygenated blood through the rest of the body.

Like most molluscs, cephalopods use hemocyanin, a copper-containing protein, rather than hemoglobin, to transport oxygen. As a result, their blood is colorless when deoxygenated and turns blue when exposed to air.

Respiration

Cephalopods exchange gases with the seawater by forcing water through their gills, which are attached to the roof of the organism. Water enters the mantle cavity on the outside of the gills, and the entrance of the mantle cavity closes. When the mantle contracts, water is forced through the gills, which lie between the mantle cavity and the funnel. The water's expulsion through the funnel can be used to power jet propulsion. The gills, which are much more efficient than those of other molluscs, are attached to the ventral surface of the mantle cavity. There is a trade-off with gill size regarding lifestyle. To achieve fast speeds, gills need to be small – water will be passed through them quickly when energy is needed, compensating for their small size. However, organisms which spend most of their time moving slowly along the bottom do not naturally pass much water through their cavity for locomotion; thus they have larger gills, along with complex systems to ensure that water is constantly washing through their gills, even when the organism is stationary. The water flow is controlled by contractions of the radial and circular mantle cavity muscles.

The gills of cephalopods are supported by a skeleton of robust fibrous proteins; the lack of mucopolysaccharides distinguishes this matrix from cartilage. The gills are also thought to be involved in excretion, with NH_4^+ being swapped with K^+ from the seawater.

Locomotion and Buoyancy

Octopuses swim headfirst, with arms trailing behind

While most cephalopods can move by jet propulsion, this is a very energy-consuming way to travel compared to the tail propulsion used by fish. The efficiency of a propellor-driven waterjet (i.e. Froude efficiency) is a more efficient model than rocket efficiency. The relative efficiency of jet propulsion decreases further as animal size increases; paralarvae are far more efficient than juvenile and adult individuals. Since the Paleozoic era, as competition with fish produced an environment where efficient motion was crucial to survival, jet propulsion has taken a back role, with fins and tentacles used to maintain a steady velocity. Whilst jet propulsion is never the sole mode of locomotion, the stop-start motion provided by the jets continues to be useful for providing bursts of high speed – not least when capturing prey or avoiding predators. Indeed, it makes cephalopods the fastest marine invertebrates, and they can out-accelerate most fish. The jet is supplemented with fin motion; in the squid, the fins flap each time that a jet is released, amplifying the thrust; they are then extended between jets (presumably to avoid sinking). Oxygenated water is taken into the mantle cavity to the gills and through muscular contraction of this cavity, the spent water is expelled through the hyponome, created by a fold in the mantle. The size difference between the posterior and anterior ends of this organ control the speed of the jet the organism can produce. The velocity of the organism can be accurately predicted for a given mass and morphology of animal. Motion of the cephalopods is usually backward as water is forced out anteriorly through the hyponome, but direction can be controlled somewhat by pointing it in different directions. Some cephalopods accompany this expulsion of water with a gunshot-like popping noise, thought to function to frighten away potential predators.

Cephalopods employ a similar method of propulsion despite their increasing size (as they grow) changing the dynamics of the water in which they find themselves. Thus their paralarvae do not extensively use their fins (which are less efficient at low Reynolds numbers) and primarily use their jets to propel themselves upwards, whereas large adult cephalopods tend to swim less efficiently and with more reliance on their fins.

Nautilus belauensis seen from the front, showing the opening of the hyponome

Early cephalopods are thought to have produced jets by drawing their body into their shells, as *Nautilus* does today. *Nautilus* is also capable of creating a jet by undulations of its funnel; this slower flow of water is more suited to the extraction of oxygen from the water. The jet velocity in *Nautilus* is much slower than in coleoids, but less musculature and energy is involved in its production. Jet thrust in cephalopods is controlled primarily by the maximum diameter of the funnel orifice (or, perhaps, the average diameter of the funnel) and the diameter of the mantle cavity. Changes in the size of the orifice are used most at intermediate velocities. The absolute velocity achieved is limited by the cephalopod's requirement to inhale water for expulsion; this intake limits the maximum velocity to eight body-lengths per second, a speed which most cephalopods can attain after two funnel-blows. Water refills the cavity by entering not only through the orifices, but also through the funnel. Squid can expel up to 94% of the fluid within their cavity in a single jet thrust. To accommodate the rapid changes in water intake and expulsion, the orifices are highly flexible and can change their size by a factor of twenty; the funnel radius, conversely, changes only by a factor of around 1.5.

Some octopus species are also able to walk along the sea bed. Squids and cuttlefish can move short distances in any direction by rippling of a flap of muscle around the mantle.

While most cephalopods float (i.e. are neutrally buoyant or nearly so; in fact most cephalopods are about 2–3% denser than seawater), they achieve this in different ways. Some, such as *Nautilus*, allow gas to diffuse into the gap between the mantle and the shell; others allow purer water to ooze from their kidneys, forcing out denser salt water from the body cavity; others, like some fish, accumulate oils in the liver; and some octopuses have a gelatinous body with lighter chlorine ions replacing sulfate in the body chemistry.

The *Macrotritopus defilippi*, or the sand-dwelling octopus, was seen mimicking both the coloration and the swimming movements of the sand-dwelling flounder *Bothus lunatus* to avoid predators. The octopuses were able to flatten their bodies and put their arms back to appear the same as the flounders as well as move with the same speed and movements.

Shell

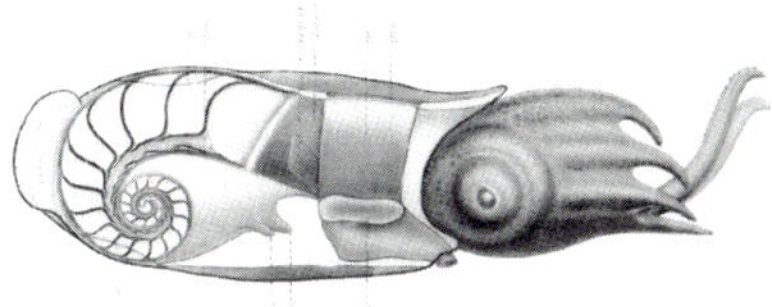

Cross section of *Spirula spirula*, showing the position of the shell inside the mantle

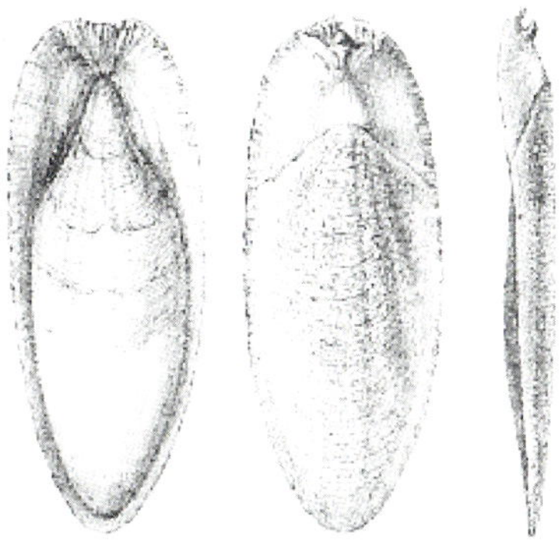

Cuttlebone of *Sepia officinalis*

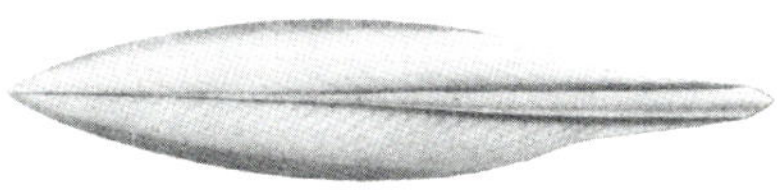

Gladius of *Sepioteuthis lessoniana*

Nautiluses are the only extant cephalopods with a true external shell. However, all molluscan shells are formed from the ectoderm (outer layer of the embryo); in cuttlefish (*Sepia* spp.), for example, an invagination of the ectoderm forms during the embryonic period, resulting in a shell (cuttlebone) that is internal in the adult. The same is true of the chitinous gladius of squid and octopuses. Cirrate octopods have arch-shaped cartilaginous fin supports, which are sometimes referred to as a "shell vestige" or "gladius". The Incirrina have either a pair of rod-shaped stylets or no vestige of an internal shell, and some squid also lack a gladius. The shelled coleoids do not form a clade or even a paraphyletic group. The *Spirula* shell begins as an organic structure, and is then very rapidly mineralized. Shells that are "lost" may be lost by resorption of the calcium carbonate component.

Females of the octopus genus *Argonauta* secrete a specialised paper-thin eggcase in which they reside, and this is popularly regarded as a "shell", although it is not attached to the body of the animal.

The largest group of shelled cephalopods, the ammonites, are extinct, but their shells are very common as fossils.

The deposition of carbonate, leading to a mineralized shell, appears to be related to the acidity of the organic shell matrix; shell-forming cephalopods have an acidic matrix, whereas the gladius of squid has a basic matrix.

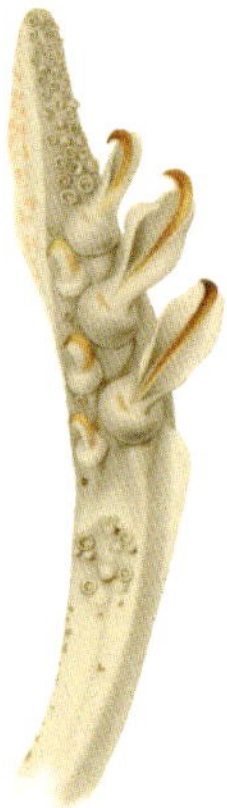

Left: A giant squid found in Logy Bay, Newfoundland. The two long feeding tentacles are visible on the extreme left and right. Right: Detail of the tentacular club of *Abraliopsis morisi*

Head Appendages

Cephalopods, as the name implies, have muscular appendages extending from their heads and surrounding their mouths. These are used in feeding, mobility, and even reproduction. In coleoids they number eight or ten. Decapods such as cuttlefish and squid have five pairs. The longer two, termed *tentacles*, are actively involved in capturing prey; they can lengthen rapidly (in as little as 15 milliseconds). In giant squid they may reach a length of 8 metres. They may terminate in a broadened, sucker-coated club. The shorter four pairs are termed *arms*, and are involved in holding and manipulating the captured organism. They too have suckers, on the side closest to the mouth; these help to hold onto the prey. Octopods only have four pairs of sucker-coated arms, as the name suggests, though developmental abnormalities can modify the number of arms expressed.

The tentacle consists of a thick central nerve cord (which must be thick to allow each sucker to be controlled independently) surrounded by circular and radial muscles. Because the volume of the tentacle remains constant, contracting the circular muscles decreases the radius and permits the rapid increase in length. Typically a 70% lengthening is achieved by decreasing the width by 23%. The shorter arms lack this capability.

The size of the tentacle is related to the size of the buccal cavity; larger, stronger tentacles can hold prey as small bites are taken from it; with more numerous, smaller tentacles, prey is swallowed whole, so the mouth cavity must be larger.

Externally shelled nautilids (*Nautilus* and *Allonautilus*) have on the order of 90 finger-like appendages, termed *tentacles*, which lack suckers but are sticky instead, and are partly retractable.

Feeding

The two-part beak of the giant squid, *Architeuthis* sp.

All living cephalopods have a two-part beak; most have a radula, although it is reduced in most octopus and absent altogether in *Spirula*. They feed by capturing prey with their tentacles, drawing it into their mouth and taking bites from it. They have a mixture of toxic digestive juices, some of which are manufactured by symbiotic algae, which they eject from their salivary glands onto their captured prey held in their mouth. These juices separate the flesh of their prey from the bone or shell. The salivary gland has a small tooth at its end which can be poked into an organism to digest it from within.

The digestive gland itself is rather short. It has four elements, with food passing through the crop, stomach and caecum before entering the intestine. Most digestion, as well as the absorption of nutrients, occurs in the digestive gland, sometimes called the liver. Nutrients and waste materials are exchanged between the gut and the digestive gland through a pair of connections linking the gland to the junction of the stomach and caecum. Cells in the digestive gland directly release pigmented excretory chemicals into the lumen of the gut, which are then bound with mucus passed through the anus as long dark strings, ejected with the aid of exhaled water from the funnel. Cephalopods tend to concentrate ingested heavy metals in their body tissue.

Radula

Amphioctopus marginatus eating a crab

The cephalopod radula consists of multiple symmetrical rows of up to nine teeth – thirteen in fossil classes. The organ is reduced or even vestigial in certain octopus species and is absent in *Spirula*. The teeth may be homodont (i.e. similar in form across a row), heterodont (otherwise), or ctenodont (comb-like). Their height, width and number of cusps is variable between species. The pattern of teeth repeats, but each row may not be identical to the last; in the octopus, for instance, the sequence repeats every five rows.

Cephalopod radulae are known from fossil deposits dating back to the Ordovician. They are usually preserved within the cephalopod's body chamber, commonly in conjunction with the mandibles; but this need not always be the case; many radulae are preserved in a range of settings in the Mason Creek. Radulae are usually difficult to detect, even when they are preserved in fossils, as the rock must weather and crack in exactly the right fashion to expose them; for instance, radulae have only been found in nine of the 43 ammonite genera, and they are rarer still in non-ammonoid forms: only three pre-Mesozoic species possess one.

Excretory System

Most cephalopods possess a single pair of large nephridia. Filtered nitrogenous waste is produced in the pericardial cavity of the branchial hearts, each of which is connected to a nephridium by a narrow canal. The canal delivers the excreta to a bladder-like renal sac, and also resorbs excess water from the filtrate. Several outgrowths of the lateral vena cava project into the renal sac, continuously inflating and deflating as the branchial hearts beat. This action helps to pump the secreted waste into the sacs, to be released into the mantle cavity through a pore.

Nautilus, unusually, possesses four nephridia, none of which are connected to the pericardial cavities.

The incorporation of ammonia is important for shell formation in terrestrial molluscs and other non-molluscan lineages. Because protein (i.e. flesh) is a major constituent of the cephalopod diet, large amounts of ammonium are produced as waste. The main organs involved with the release of this excess ammonium are the gills. The rate of release is lowest in the shelled cephalopods *Nautilus* and *Sepia* as a result of their using nitrogen to fill their shells with gas to increase buoyancy. Other cephalopods use ammonium in a similar way, storing the ions (as ammonium chloride) to reduce their overall density and increase buoyancy.

Reproduction and Life Cycle

Female *Argonauta argo* with eggcase and eggs

Cephalopods are a diverse group of species, but share common life history traits, for example they have a rapid growth rate and short life spans. Stearns (1992) suggested that in order to produce the largest possible number of viable offspring, spawning events depend on ecological environmental factors of the organism. The majority of cephalopods do not provide parental care to their offspring, except for example, octopus, which helps this organism increase the survival rate of their offspring. Marine species life cycles are affected by various environmental conditions. The development of a cephalopod embryo can be greatly affected by temperature, oxygen saturation, pollution, light intensity, and salinity. These factors are important to the rate of embryonic development and the success of hatching of the embryos. Food availability also plays an important role in the reproductive cycle of cephalopods. A limitation of food influences the timing of spawning along with their function and growth. Spawning time and spawning vary among marine species; it's correlated with temperature, though cephalopods in shallow water spawn in cold months so that the offspring would hatch at warmer temperatures. Breeding can last from several days to a month.

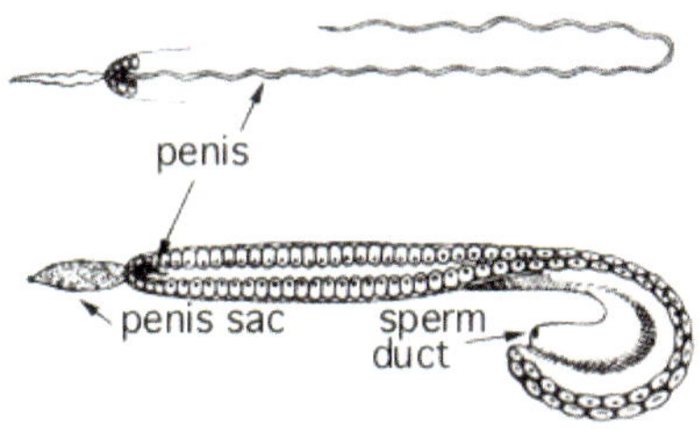

Detail of the hectocotylus of Ocythoe tuberculata

A dissected male specimen of *Onykia ingens*, showing a non-erect penis

A specimen of the same species exhibiting elongation of the penis to 67 cm in length

Sexual Maturity

Cephalopods that are sexually mature and of adult size begin spawning and reproducing. After the transfer of genetic material to the following generation, the adult cephalopods then die. Sexual maturation in male and female cephalopods can be observed internally by the enlargement of gonads and accessory glands. Mating would be a poor indicator of sexual maturation in females; they can receive sperm when not fully reproductively mature and store them until they are ready to fertilize the eggs. Males are more aggressive in their pre-mating competition when in the presence of immature females than when competing for a sexually mature female. Most cephalopod males develop a hectocotylus, an arm tip which is capable of transferring their spermatozoa into the female mantel cavity. Though not all species use a hectocotylus; for example, the adult nautilus releases a spadix. An indication of sexual maturity of females is the development of brachial photophores to attract mates.

Fertilization

Cephalopods are not broadcast spawners. During the process of fertilization, the females use sperm provided by the male via external fertilization. Internal fertilization is seen only in octopods. The initiation of copulation begins when the male catches a female and wraps his arm around her, either in a "male to female neck" position or mouth to mouth position, depending on the species. The males then initiate the process of fertilization by contracting their mantle several times to release the spermatozoa. Cephalopods often mate several times, which influences males to mate longer with females that have previously, nearly tripling the amount of contractions of the mantle. To ensure the fertilization of the eggs, female cephalopods release a sperm-attracting peptide through the gelatinous layers of the egg to direct the spermatozoa. Female cephalopods lay eggs in clutches; each egg is composed of a protective coat to ensure the safety of the developing embryo when released into the water column. Reproductive strategies differ between cephalopod species. In giant pacific octopus, large eggs are laid in a den; it will often take several days to lay all of them. Once the eggs are released and attached to a sheltered substrate the females then die. In some species of cephalopods, egg clutches are anchored to substrates by a mucilaginous adhesive substance. These eggs are swelled with preivitelline fluid (PVF), a hypertonic fluid that prevents premature hatching. Fertilized egg clusters are neutrally buoyant depending at the depth that they were laid but can also be found in substrates such as sand, matrix of corals, or seaweed. Because these species do not provide parental care for their offspring, egg capsules can be injected with ink by the female in order to camouflage the embryos from predators.

Male-male Competition

Most cephalopods engage in aggressive sex: a protein in the male capsule sheath stimulates this behavior. They also engage in male-male aggression, where larger males tend to win the interactions. When a female is near, the males charge one another continuously and flail their arms. If neither male backs away, the arms extend to the back, exposing the mouth, followed by the biting of arm tips. During mate competition males also participate in a technique called flushing. This technique is used by the second male attempting to mate with a female. Flushing removes spermatophores in the buccal cavity that was placed there by the first mate by forcing water into the cavity. Another behavior that males engage in is sneaker mating or mimicry- smaller males adjust their behavior to that of a female in order to reduce aggression. By using this technique, they are able to fertilize the eggs while the larger male is distracted by different male. During this process, the sneaker males quickly insert drop like sperm into the seminal receptacle.

Mate Choice

Mate choice is seen in cuttlefish species, where females prefer some males over others, though characteristics of the preferred males are unknown. A hypothesis states that females reject males by olfactory cues rather than visual cues. Several cephalopod species are polyandrous- accepting and storing multiple male spermatophores, which has been identified by DNA fingerprinting. Females are no longer receptive to mating attempts when holding their eggs in their arm. Females can store sperm in two places (1) the buccal cavity where recently mated males place their spermatophores, and (2) the internal sperm-storage receptacles where sperm packages from previous males are stored. Spermatophore storage results in sperm competition; which states that the female controls which mate fertilizes the eggs. In order to reduce this sort of competition, males develop agonistic behaviors like mate guarding and flushing. The *Hapalochlaena lunulata*, or the blue-ringed octopus, is unable to distinguish between males and females and readily mates with both.

Sexual Dimorphism

In a variety of marine organisms it is seen that females are larger in size compared to the males in some close related species. In some lineages, such as the blanket octopus, males become structurally smaller and smaller resembling a term, “dwarfism” dwarf males usually occurs at low densities. The blanket octopus male is an example of sexual-evolutionary dwarfism; females grow 10,000 to 40,000 times larger than the males and the sex ratio between males and females can be distinguished right after hatching of the eggs.

Egg cases laid by a female squid

Embryology

Cephalopod eggs span a large range of sizes, from 1 to 30 mm in diameter. The fertilised ovum initially divides to produce a disc of germinal cells at one pole, with the yolk remaining at the opposite pole. The germinal disc grows to envelop and eventually absorb the yolk, forming the embryo. The tentacles and arms first appear at the hind part of the body, where the foot would be in other molluscs, and only later migrate towards the head.

The funnel of cephalopods develops on the top of their head, whereas the mouth develops on the opposite surface. The early embryological stages are reminiscent of ancestral gastropods and extant Monoplacophora.

The shells develop from the ectoderm as an organic framework which is subsequently mineralised. In *Sepia*, which has an internal shell, the ectoderm forms an invagination whose pore is sealed off before this organic framework is deposited.

Development

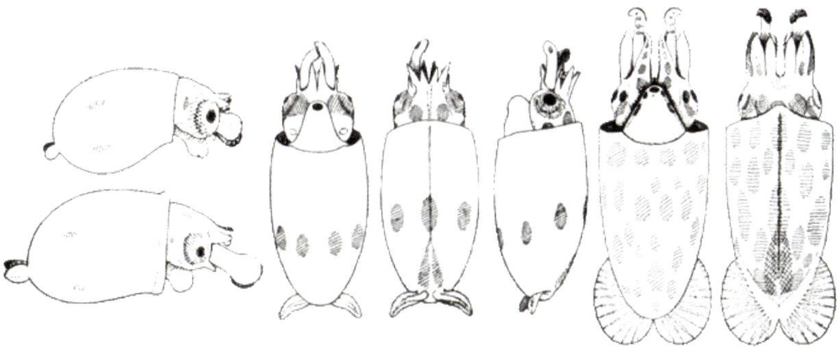

Chtenopteryx sicula paralarvae. Left: Two very young paralarvae. The circular tentacular clubs bear approximately 20 irregularly arranged suckers. Two chromatophores are present on each side of the mantle. Centre: Ventral, dorsal and side views of a more advanced paralarva. An equatorial circulet of seven large yellow-brown chromatophores is present on the mantle. Posteriorly the expanded vanes of the gladius are visible in the dorsal view. Right: Ventral and dorsal views of a very advanced paralarva.

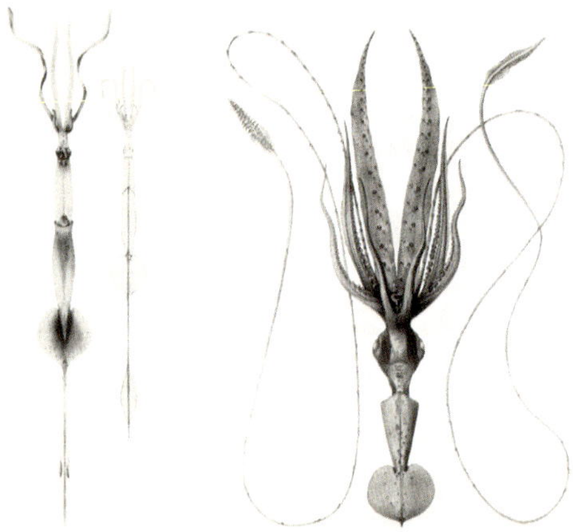

Left: Immature specimens of *Chiroteuthis veranyi*. In this paralarval form, known as the doratopsis stage, the pen is longer than the mantle and 'neck' combined Right: A mature *Chiroteuthis veranyi*. This species has some of the longest tentacles in proportion to its size of any known cephalopod.

The length of time before hatching is highly variable; smaller eggs in warmer waters are the fastest to hatch, and newborns can emerge after as little as a few days. Larger eggs in colder waters can develop for over a year before hatching.

The process from spawning to hatching follows a similar trajectory in all species, the main variable being the amount of yolk available to the young and when it is absorbed by the embryo.

Unlike most other molluscs, cephalopods do not have a morphologically distinct larval stage. Instead the juveniles are known as paralarvae. They quickly learn how to hunt, using encounters with prey to refine their strategies.

Growth in juveniles is usually allometric, whilst adult growth is isometric.

Evolution

The traditional view of cephalopod evolution holds that they evolved in the Late Cambrian from a monoplacophoran-like ancestor with a curved, tapering shell, which was closely related to the gastropods (snails). The similarity of the early shelled cephalopod *Plectronoceras* to some gastropods was used in support of this view. The development of a siphuncle would have allowed the shells of these early forms to become gas-filled (thus buoyant) in order to support them and keep the shells upright while the animal crawled along the floor, and separated the true cephalopods from putative ancestors such as *Knightoconus*, which lacked a siphuncle. Neutral or positive buoyancy (i.e. the ability to float) would have come later, followed by swimming in the Plectronocerida and eventually jet propulsion in more derived cephalopods.

However, some morphological evidence is difficult to reconcile with this view, and the redescription of *Nectocaris pteryx*, which did not have a shell and appeared to possess jet propulsion in the manner of "derived" cephalopods, complicated the question of the order in which cephalopod features developed – provided *Nectocaris* is a cephalopod at all.

Early cephalopods were likely predators near the top of the food chain. After the late Cambrian extinction led to the disappearance of many Anomalocaridids, predatory niches became available for other animals. During the Ordovician period the primitive cephalopods underwent pulses of diversification to become diverse and dominant in the Paleozoic and Mesozoic seas.

In the Early Palaeozoic, their range was far more restricted than today; they were mainly constrained to sublittoral regions of shallow shelves of the low latitudes, and usually occur in association with thrombolites. A more pelagic habit was gradually adopted as the Ordovician progressed. Deep-water cephalopods, whilst rare, have been found in the Lower Ordovician – but only in high-latitude waters. The mid Ordovician saw the first cephalopods with septa strong enough to cope with the pressures associated with deeper water, and could inhabit depths greater than 100–200 m. The direction of shell coiling would prove to be crucial to the future success of the lineages; endogastric coiling would only permit large size to be attained with a straight shell, whereas exogastric coiling – initially rather rare – permitted the spirals familiar from the fossil record to develop, with their corresponding large size and diversity. (Endogastric mean the shell is curved so as the ventral or lower side is longitudinally concave (belly in); exogastric means the shell is curved so as the ventral side is longitudinally convex (belly out) allowing the funnel to be pointed backwards beneath the shell.)

An ammonoid with the body chamber missing, showing the septal surface (especially at right) with its undulating lobes and saddles

The ancestors of coleoids (including most modern cephalopods) and the ancestors of the modern nautilus, had diverged by the Floian Age of the Early Ordovician Period, over 470 million years ago. The Bactritida, a Silurian–Triassic group of orthocones, are widely held to be paraphyletic to the coleoids and ammonoids, that is, the latter groups arose from within the Bactritida. An increase in the diversity of the coleoids and ammonoids is observed around the start of the Devonian period, and corresponds with a profound increase in fish diversity. This could represent the origin of the two derived groups.

Unlike most modern cephalopods, most ancient varieties had protective shells. These shells at first were conical but later developed into curved nautiloid shapes seen in modern nautilus species. Competitive pressure from fish is thought to have forced the shelled forms into deeper water, which provided an evolutionary pressure towards shell loss and gave rise to the modern coleoids, a change which led to greater metabolic costs associated with the loss of buoyancy, but which allowed them to recolonise shallow waters. However, some of the straight-shelled nautiloids evolved into belemnites, out of which some evolved into squid and cuttlefish. The loss of the shell may also have resulted from evolutionary pressure to increase manoeuvrability, resulting in a more fish-like habit.

There has been debate on the embryological origin of cephalopod appendages. Until the mid-twentieth century, the "Arms as Head" hypothesis was widely recognized. In this theory, the arms and tentacles of cephalopods look similar to the head appendages of gastropods, suggesting that they might be homologous structures. Cephalopod appendages surround the mouth, so logically they could be derived from embryonic head tissues. However, the "Arms as Foot" hypothesis, proposed by Adolf Naef in 1928, has increasingly been favoured; for example, fate mapping of limb buds in the chambered nautilus indicates that limb buds originate from "foot" embryonic tissues.

Taxonomy

Chambered nautilus (*Nautilus pompilius*)

Common cuttlefish (*Sepia officinalis*)

Atlantic bobtail (*Sepiola atlantica*)

European squid (*Loligo vulgaris*)

Common octopus (*Octopus vulgaris*)

The classification presented here, for recent cephalopods, follows largely from Current Classification of Recent Cephalopoda (May 2001), for fossil cephalopods takes from Arkell et al. 1957, Teichert and Moore 1964, Teichert 1988, and others. The three subclasses are traditional, corresponding to the three orders of cephalopods recognized by Bather.

Class Cephalopoda († indicates extinct groups)

- Subclass Nautiloidea: Fundamental ectocochleate cephalopods that provided the source for the Ammonoidea and Coleoidea.
  - Order † Plectronocerida: the ancestral cephalopods from the Cambrian Period
  - Order † Ellesmerocerida (500 to 470 Ma)
  - Order † Endocerida (485 to 430 Ma)
  - Order † Actinocerida (480 to 312 Ma)
  - Order † Discosorida (482 to 392 Ma)
  - Order † Pseudorthocerida (432 to 272 Ma)
  - Order † Tarphycerida (485 to 386 Ma)
  - Order † Oncocerida (478.5 to 324 Ma)
  - Order Nautilida (extant; 410.5 Ma to present)
  - Order † Orthocerida (482.5 to 211.5 Ma)
  - Order † Ascocerida (478 to 412 Ma)
  - Order † Bactritida (418.1 to 260.5 Ma)
- Subclass † Ammonoidea: Ammonites (479 to 66 Ma)
  - Order † Goniatitida (388.5 to 252 Ma)

- Order † Ceratitida (254 to 200 Ma)
- Order † Ammonitida (215 to 66 Ma)

- Subclass Coleoidea (410.0 Ma-Rec)
  - Cohort Neocoleoidea
    - Superorder Decapodiformes (also known as Decabrachia or Decembranchiata)
      - † Belemnoidea: Belemnites and kin
    - Genus † *Jeletzkya*
    - Order † Aulacocerida (265 to 183 Ma)
    - Order † Phragmoteuthida (189.6 to 183 Ma)
    - Order † Hematitida (339.4 to 318.1 Ma)
    - Order † Belemnitida (339.4 to 66 Ma)
    - Genus † *Belemnoteuthis* (189.6 to 183 Ma)
      - ?Order † Boletzkyida
      - Order Spirulida: Ram's horn squid
      - Order Sepiida: cuttlefish
      - Order Sepiolida: pygmy, bobtail and bottletail squid
      - Order Teuthida: squid
    - Superorder Octopodiformes (also known as Vampyropoda)
      - Family † Trachyteuthididae
      - Order Vampyromorphida: Vampire squid
      - Order Octopoda: octopus

Other classifications differ, primarily in how the various decapod orders are related, and whether they should be orders or families.

Suprafamilial Classification of the Treatise

This is the older classification that combines those found in parts K and L of the *Treatise on Invertebrate Paleontology*, which forms the basis for and is retained in large part by classifications that have come later.

Nautiloids in general (Teichert and Moore, 1964) sequence as given.

Subclass † Endoceratoidea. Not used by Flower, e.g. Flower and Kummel 1950, interjocerids included in the Endocerida.

Order † Endocerida

Order † Intejocerida

Subclass † Actinoceratoidea Not used by Flower, ibid

Order † Actinocerida

Subclass † Nautiloidea Nautiloidea in the restricted sense.

Order † Ellesmerocerida Plectronocerida subsequently split off as separate order.

Order † Orthocerida Includes orthocerids and pseudorthocerids

Order † Ascocerida

Order † Oncocerida

Order † Discosorida

Order † Tarphycerida

Order † Barrandeocerida A polyphyletic group now included in the Tarphycerida

Order Nautilida

Subclass † Bactritoidea

Order † Bactritida

Paleozoic Ammonoidea (Miller, Furnish and Schindewolf, 1957)

Suborder † Anarcestina

Suborder † Clymeniina

Suborder † Goniatitina

Suborder † Prolecanitina

Mesozoic Ammonoidea (Arkel et al., 1957)

Suborder † Ceratitina

Suborder † Phylloceratina

Suborder † Lytoceratina

Suborder † Ammonitina

Subsequent revisions include the establishment of three Upper Cambrian orders, the Plectronocerida, Protactinocerida and Yanhecerida; separation of the pseudorthocerids as the Pseudorthocerida, and elevating orthoceritoids as the Subclass Orthoceratoidea.

Shevyrev classification

Shevyrev (2005) suggested a division into eight subclasses, mostly comprising the more diverse and numerous fossil forms, although this classification has been criticized as arbitrary.

Holotype of *Ostenoteuthis siroi* from family Ostenoteuthidae.

Various species of ammonites

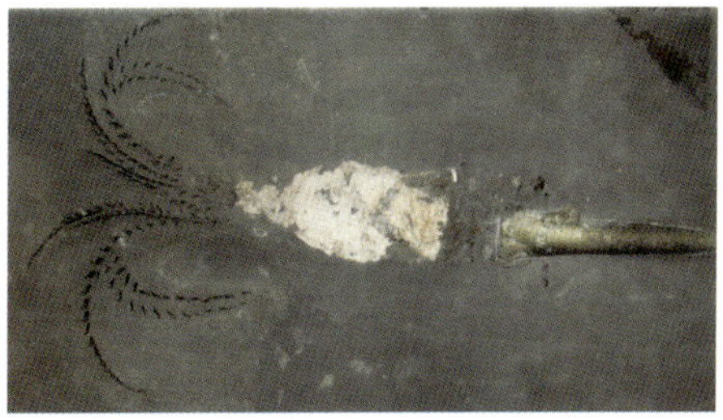

A fossilised belemnite

Class Cephalopoda

- Subclass † Ellesmeroceratoidea
  - Order † Plectronocerida (501 to 490 Ma)
  - Order † Protactinocerida
  - Order † Yanhecerida
  - Order † Ellesmerocerida (500 to 470 Ma)
- Subclass † Endoceratoidea (485 to 430 Ma)
  - Order † Endocerida (485 to 430 Ma)
  - Order † Intejocerida (485 to 480 Ma)
- Subclass † Actinoceratoidea
  - Order † Actinocerida (480 to 312 Ma)
- Subclass Nautiloidea (490.0 Ma- Rec)
  - Order † Basslerocerida (490 to 480 Ma)
  - Order † Tarphycerida (485 to 386 Ma)
  - Order † Lituitida (485 to 480 Ma)
  - Order † Discosorida (482 to 392 Ma)
  - Order † Oncocerida (478.5 to 324 Ma)
  - Order Nautilida (410.5 Ma-Rec)
- Subclass † Orthoceratoidea (482.5 to 211.5 Ma)
  - Order † Orthocerida (482.5 to 211.5 Ma)
  - Order † Ascocerida (478 to 412 Ma)
  - Order † Dissidocerida (479 to 457.5 Ma)

    - Order † Bajkalocerida
- Subclass † Bactritoidea (422 to 252 Ma)
- Subclass † Ammonoidea (410 to 66 Ma)
- Subclass Coleoidea (410.0 Ma-rec)

Cladistic Classification

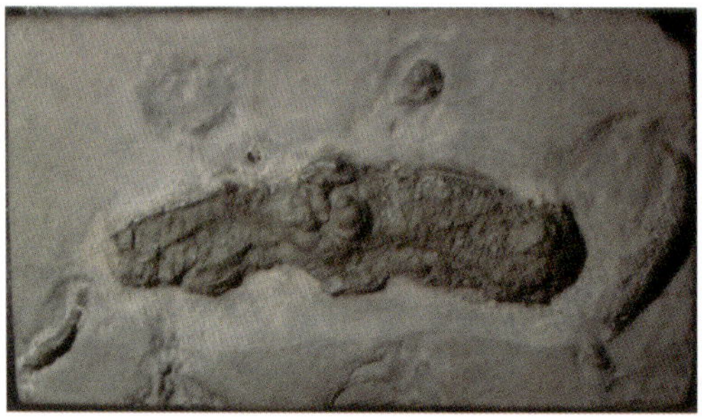

Pyritized fossil of *Vampyronassa rhodanica,* a vampyromorphid from the Lower Callovian (166.1 million years ago)

Another recent system divides all cephalopods into two clades. One includes nautilus and most fossil nautiloids. The other clade (Neocephalopoda or Angusteradulata) is closer to modern coleoids, and includes belemnoids, ammonoids, and many orthocerid families. There are also stem group cephalopods of the traditional Ellesmerocerida that belong to neither clade.

The coleoids, despite some doubts, appear from molecular data to be monophyletic.

In Culture

Pen and wash drawing of an imagined colossal octopus attacking a ship, by the malacologist Pierre de Montfort, 1801

Ancient seafaring people were aware of cephalopods, as evidenced by artworks such as a stone carving found in the archaeological recovery from Bronze Age Minoan Crete at Knossos (1900 – 1100 BC) has a depiction of a fisherman carrying an octopus. The terrifyingly powerful Gorgon of Greek mythology may have been inspired by the octopus or squid, the octopus's body representing the severed head of Medusa, the beak as the protruding tongue and fangs, and its tentacles as the snakes.

The NROL-39 mission patch, depicting the National Reconnaissance Office as an octopus with a long reach

The Kraken are legendary sea monsters of giant proportions said to dwell off the coasts of Norway and Greenland, usually portrayed in art as giant cephalopods attacking ships. Linnaeus included it in the first edition of his 1735 *Systema Naturae*. A Hawaiian creation myth says that the present cosmos is the last of a series which arose in stages from the ruins of the previous universe. In this account, the octopus is the lone survivor of the previous, alien universe. The Akkorokamui is a gigantic tentacled monster from Ainu folklore.

A battle with an octopus plays a significant role in Victor Hugo's book *Travailleurs de la mer* (*Toilers of the Sea*), relating to his time in exile on Guernsey. Ian Fleming's 1966 short story collection *Octopussy and The Living Daylights*, and the 1983 *James Bond* film were partly inspired by Hugo's book.

Japanese erotic art, *shunga*, includes ukiyo-e woodblock prints such as Katsushika Hokusai's 1814 print *Tako to ama* (The Dream of the Fisherman's Wife), in which an ama diver is sexually intertwined with a large and a small octopus. The print is a forerunner of tentacle erotica. The biologist P. Z. Myers noted in his science blog, Pharyngula, that octopuses appear in "extraordinary" graphic illustrations involving women, tentacles, and bare breasts.

Since it has numerous arms emanating from a common centre, the octopus is often used as a symbol for a powerful and manipulative organisation, usually negatively.

References

- Cobb, Matthew (2000-09-01). "Reading and writing The Book of Nature: Jan Swammerdam (1637–1680)". Endeavour. 24 (3): 122–128. doi:10.1016/S0160-9327(00)01306-5
- Schmidt, Günther (1993). Giftige und gefährliche Spinnentiere [Poisonous and dangerous arachnids] (in German). Westarp Wissenschaften. p. 75. ISBN 3-89432-405-8
- Rice C.M. et al. (1995). «A Devonian auriferous hot spring system, Rhynie, Scotland". Journal of the Geological Society, London 152: 229–250. doi:10.1144/gsjgs.152.2.0229
- Egerton, Frank N. (2008-10-01). "A History of the Ecological Sciences, Part 30: Invertebrate Zoology and Parasitology During the 1700s". The Bulletin of the Ecological Society of America. 89 (4): 407–433. doi:10.1890/0012-9623(2008)89[407:AHOTES]2.0.CO;2. ISSN 2327-6096

- Shultz, Jeffrey W. (January 1, 1992). "Muscle Firing Patterns in Two Arachnids Using Different Methods of Propulsive Leg Extension". Journal of Experimental Biology. 162 (1): 313–329. ISSN 1477-9145. Retrieved 2012-05-19
- Cees H. J. Hof (1998). "Fossil stomatopods (Crustacea: Malacostraca) and their phylogenetic impact". Journal of Natural History. 32 (10 & 11): 1567–1576. doi:10.1080/00222939800771101
- Engel, Michael S.; David A. Grimaldi (2004). "New light shed on the oldest insect". Nature 427 (6975): 627–630. doi:10.1038/nature02291. PMID 14961119
- Ruppert, E.E.; Fox, R.S. & Barnes, R.D. (2004). "Mollusca". Invertebrate Zoology (7th ed.). Brooks / Cole. pp. 290–291. ISBN 0030259827
- Alouini Z., Jemli M. (2001). "Destruction of helminth eggs by photosensitized porphyrin". Journal of Environmental Monitoring. 3: 548–551. CiteSeerX 10.1.1.691.4034 . doi:10.1039/b103471p
- Richter, Stefan; Meier, Rudolf (1994-01-01). "The Development of Phylogenetic Concepts in Hennig's Early Theoretical Publications (1947-1966)". Systematic Biology. 43 (2): 212–221. doi:10.2307/2413462
- "Crustaceans — bugs of the sea". Te Ara: The Encyclopedia of New Zealand. Ministry for Culture and Heritage. Retrieved 2016-09-10
- Tompa, A. S. (December 1976). "A comparative study of the ultrastructure and mineralogy of calcified land snail eggs (Pulmonata: Stylommatophora)". Journal of Morphology. 150 (4): 861–887. doi:10.1002/jmor.1051500406. hdl:2027.42/50263
- Novotny, Vojtech et al 2002.. "Low host specificity of herbivorous insects in a tropical forest". Nature 416 (6883): 841–844. doi:10.1038/416841a. PMID 11976681
- Ruppert, E.; Fox, R. & Barnes, R. (2007). Invertebrate Zoology: A Functional Evolutionary Approach (7th ed.). Thomson Learning. ISBN 0-03-025982-7
- Jensen, K. R. (February 1993). "Morphological adaptations and plasticity of radular teeth of the Sacoglossa (= Ascoglossa) (Mollusca: Opisthobranchia) in relation to their food plants". Biological Journal of the Linnean Society. 48 (2): 135–155. doi:10.1111/j.1095-8312.1993.tb00883.x
- Shultz, Jeffrey W. (February 6, 2005). "Evolution of locomotion in arachnida: The hydraulic pressure pump of the giant whipscorpion, Mastigoproctus giganteus (Uropygi)". Journal of Morphology. 210 (1): 13–31. doi:10.1002/jmor.1052100103. ISSN 1097-4687
- Butterfield, N.J. (May 2008). "An Early Cambrian Radula". Journal of Paleontology. 82 (3): 543–554. doi:10.1666/07-066.1
- "Schistosomiasis Research Group, University of Cambridge, UK". Archived from the original on 13 October 2014. Retrieved 19 December 2014
- Sensenig, Andrew T. & Shultz, Jeffrey W. (2004). "Elastic energy storage in the pedipedal joints of scorpions and sun-spiders (Arachnida, Scorpiones, Solifugae)". Journal of Arachnology. 32 (1): 1–10. doi:10.1636/S02-73. ISSN 0161-8202
- Zong-Jie, F. (2006). "An introduction to Ordovician bivalves of southern China, with a discussion of the early evolution of the Bivalvia". Geological Journal. 41(3–4): 303–328. doi:10.1002/gj.1048

Chapter 4

Ethology: Animal Behavior

Animal behavior is studied under the domain of ethology. In order to completely understand animal behavior, it is necessary to understand the processes related to it. The following chapter elucidates the topics such as animal communication, consciousness, emotions, etc.

Ethology is a branch of zoology concerned with the study of animal behavior. Ethologists take a comparative approach, studying behaviors ranging from kinship, cooperation, and parental investment, to conflict, sexual selection, and aggression across a variety of species. Today *ethology* as a disciplinary label has largely been replaced by behavioral ecology and evolutionary psychology. These rapidly growing fields tend to place greater emphasis on social relationships rather than on the individual animal; however, they retain ethology's tradition of fieldwork and its grounding in evolutionary theory.

The egg-rolling behavior of the greylag goose is a widely cited example of a *fixed-action pattern,* one of the key concepts used by ethologists to explain animal behavior

The study of animal behavior touches upon the fact that people receive joy from nature and also typically see themselves in a special role as stewards of creation. Behavior is one aspect of the vast diversity of nature that enhances human enjoyment. People are fascinated with the many behaviors of animals, whether the communication "dance" of honeybees, or the hunting behavior of the big cats, or the altruistic behavior of a dolphin. In addition, humans generally see themselves with the responsibility to love and care for nature.

The study of animal behavior also helps people to understand more about themselves. From an evolutionary point of view, organisms of diverse lineages are related through the process of descent with modification. From a religious point of view, human also stand as "microcosms of nature" (Burns 2006). Thus, the understanding of animals helps to better understand ourselves.

Ethologists engage in hypothesis-driven experimental investigation, often in the field. This combination of lab work with field study reflects an important conceptual underpinning of the discipline: behavior is assumed to be *adaptive*; i.e., something that makes it better suited in its environment and consequently improves its chances of survival and reproductive success.

Ethology emerged as a discrete discipline in the 1920s, through the efforts of Konrad Lorenz, Karl von Frisch, and Niko Tinbergen, who were jointly awarded the 1973 Nobel Prize in Physiology or Medicine for their contributions to the study of behavior. They were in turn influenced by the foundational work of, among others, ornithologists Oskar Heinroth and Julian Huxley and the American myrmecologist (study of ants) William Morton Wheeler, who popularized the term *ethology* in a seminal 1902 paper.

Important Concepts

One of the key ideas of classical ethology is the concept of fixed action patterns (FAPs). FAPs are stereotyped behaviors that occur in a predictable, inflexible sequence in response to an identifiable stimulus from the environment.

For example, at the sight of a displaced egg near the nest, the greylag goose *(Anser anser)* will roll the egg back to the others with its beak. If the egg is removed, the animal continues to engage in egg-rolling behavior, pulling its head back as if an imaginary egg is still being maneuvered by the underside of its beak. It will also attempt to move other egg-shaped objects, such as a golf ball, doorknob, or even an egg too large to have been laid by the goose itself.

Another important concept is *filial imprinting,* a form of learning that occurs in young animals, usually during a critical, formative period of their lives. During imprinting, a young animal learns to direct some of its social responses to a parent or sibling.

Despite its valuable contributions to the study of animal behavior, classical ethology also spawned problematic general theories that viewed even complex behaviors as genetically hardwired (i.e., *innate* or *instinctive*). Models of behavior have since been revised to account for more flexible decision-making processes.

Methodology

Tinbergen's Four Questions for Ethologists

The practice of ethological investigation is rooted in hypothesis-driven experimentation. Lorenz's collaborator, Niko Tinbergen, argued that ethologists should consider the following categories when attempting to formulate a hypothesis that explains any instance of behavior:

- Function: How does the behavior impact the animal's chance of survival and reproduction?
- Mechanism: What are the stimuli that elicit the response? How has the response been modified by recent learning?
- Development: How does the behavior change with age? What early experiences are necessary for the behavior to be demonstrated?
- Evolutionary history: How does the behavior compare with similar behavior in related

species? How might the behavior have arisen through the evolutionary development of the species, genus, or group?

The four questions are meant to be complementary, revealing various facets of the motives underlying a given behavior.

Using Fieldwork to Test Hypotheses

As an example of how an ethologist might approach a question about animal behavior, consider the study of hearing in an echolocating bat. A species of bat may use frequency chirps to probe the environment while in flight. A traditional neuroscientific study of the auditory system of the bat would involve anesthetizing it, performing a craniotomy to insert recording electrodes in its brain, and then recording neural responses to pure tone stimuli played from loudspeakers. In contrast, an ideal ethological study would attempt to replicate the natural conditions of the animal as closely as possible. It would involve recording from the animal's brain while it is awake, producing its natural calls while performing a behavior such as insect capture.

Key Principles and Concepts

Behaviors are Adaptive Responses to Natural Selection

Because ethology is understood as a branch of biology, ethologists have been particularly concerned with the evolution of behavior and the understanding of behavior in terms of the theory of natural selection. In one sense, the first modern ethologist was Charles Darwin, whose book *The Expression of the Emotions in Man and Animals* (1872) has influenced many ethologists. (Darwin's protégé George Romanes became one of the founders of comparative psychology, positing a similarity of cognitive processes and mechanisms between animals and humans.)

Note, however, that this concept is necessarily speculative. Behaviors are not found as fossils and cannot be traced through the geological strata. And concrete evidence for the theory of modification by natural selection is limited to microevolution—that is, evolution at or below the level of species. The evidence that natural selection directs changes on the macroevolutionary level necessarily involves extrapolation from these evidences on the microevolutionary level. Thus, although scientists frequently allude to a particular behavior having evolved by natural selection in response to a particular environment, this involves speculation as opposed to concrete evidence.

Animals use Fixed Action Patterns in Communication

As mentioned above, a *fixed action pattern (FAP)* is an instinctive behavioral sequence produced by a neural network known as the *innate releasing mechanism* in response to an external sensory stimulus called the *sign stimulus* or *releaser*. Once identified by ethologists, FAPs can be compared across species, allowing them to contrast similarities and differences in behavior with similarities and differences in form (morphology).

An example of how FAPs work in animal communication is the classic investigation by Austrian ethologist Karl von Frisch of the so-called "dance language" underlying bee communication. The dance is a mechanism for successful foragers to recruit members of the colony to new sources of nectar or pollen.

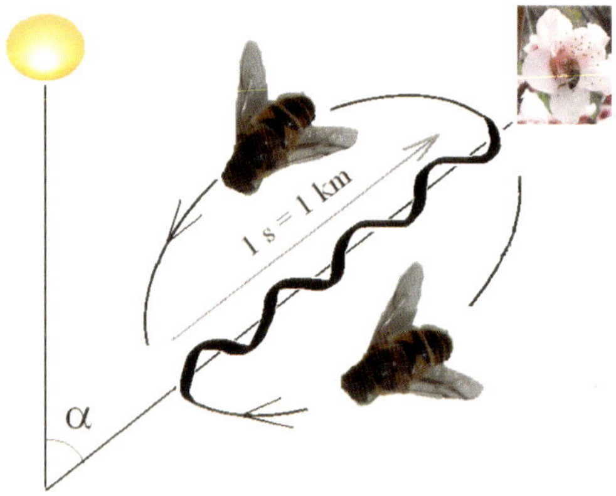

The honeybee's figure-eight dance is a fixed-action pattern that communicates information to other members of the group: the angle from the sun indicates the direction of a food source; the duration signifies its distance

Imprinting is a Type of Learning Behavior

Imprinting describes any kind of phase-sensitive learning (i.e., learning that occurs at a particular age or life stage) during which an animal learns the characteristics of some stimulus, which is therefore said to be "imprinted" onto the subject.

The best known form of imprinting is *filial imprinting,* in which a young animal learns the characteristics of its parent. Lorenz observed that the young of waterfowl such as geese spontaneously followed their mothers from almost the first day after they were hatched. Lorenz demonstrated how incubator-hatched geese would imprint on the first suitable moving stimulus they saw within what he called a critical period of about 36 hours shortly after hatching. Most famously, the goslings would imprint on Lorenz himself (more specifically, on his wading boots).

Sexual imprinting, which occurs at a later stage of development, is the process by which a young animal learns the characteristics of a desirable mate. For example, male zebra finches appear to prefer mates with the appearance of the female bird that rears them, rather than mates of their own type. *Reverse* sexual imprinting has also observed: when two individuals live in close domestic proximity during their early years, both are desensitized to later sexual attraction. This phenomenon, known as the *Westermarck effect,* has probably evolved to suppress inbreeding.

Relation to Comparative Psychology

In order to summarize the defining features of ethology, it might be helpful to compare classical ethology to early work in comparative psychology, an alternative approach to the study of animal behavior that also emerged in the early 20th century. The rivalry between these two fields stemmed in part from disciplinary politics: ethology, which had developed in Europe, failed to gain a strong foothold in North America, where comparative psychology was dominant.

Broadly speaking, comparative psychology studies general processes, while ethology focuses on adaptive specialization. The two approaches are complementary rather than competitive, but they do lead to different perspectives and sometimes to conflicts of opinion about matters of substance:

- Comparative psychology construes its study as a branch of psychology rather than as an outgrowth of biology. Thus, where comparative psychology sees the study of animal behavior in the context of what is known about human psychology, ethology situates animal behavior in the context of what is known about animal anatomy, physiology, neurobiology, and phylogenetic history.

- Comparative psychologists are interested more in similarities than differences in behavior; they are seeking general laws of behavior, especially relating to development, which can then be applied to all animal species, including humans. Hence, early comparative psychologists concentrated on gaining extensive knowledge of the behavior of a few species, while ethologists were more interested in gaining knowledge of behavior in a wide range of species in order to be able to make principled comparisons across taxonomic groups.
- Comparative psychologists focused primarily on lab experiments involving a handful of species, mainly rats and pigeons, whereas ethologists concentrated on behavior in natural situations.

Since the 1970s, however, animal behavior has become an integrated discipline, with comparative psychologists and ethological animal behaviorists working on similar problems and publishing side by side in the same journals.

Recent Developments in the Field

In 1970, the English ethologist John H. Crook published an important paper in which he distinguished *comparative ethology* from *social ethology*. He argued that the ethological studies published to date had focused on the former approach—looking at animals as individuals—whereas in the future ethologists would need to concentrate on the social behavior of animal groups.

Since the appearance of E. O. Wilson's seminal book *Sociobiology: The New Synthesis* in 1975, ethology has indeed been much more concerned with the social aspects of behavior, such as phenotypic altruism and cooperation. Research has also been driven by a more sophisticated version of evolutionary theory associated with Wilson and Richard Dawkins.

Furthermore, a substantial rapprochement with comparative psychology has occurred, so the modern scientific study of behavior offers a more or less seamless spectrum of approaches—from animal cognition to comparative psychology, ethology, and behavioral ecology. *Evolutionary psychology*, an extension of behavioral ecology, looks at commonalities of cognitive processes in humans and other animals as we might expect natural selection to have shaped them. Another promising subfield is *neuroethology*, concerned with how the structure and functioning of the brain controls behavior and makes learning possible.

Animal Behavior

Animal behaviour, the concept, broadly considered, referring to everything animals do, including movement and other activities and underlying mental processes. Human fascination with animal behaviour probably extends back millions of years, perhaps even to times before the ancestors of the species became human in the modern sense. Initially, animals were probably observed for practical reasons because early human survival depended on knowledge of animal behaviour. Whether hunting wild game, keeping domesticated animals, or escaping an attacking predator, success required intimate knowledge of an animal's habits. Even today, information about animal behaviour is of considerable importance. For example, in Britain, studies on the social organization and the ranging patterns of badgers (Meles meles) have helped reduce the spread of tuberculosis among cattle, and studies of sociality in foxes (Vulpes vulpes) assist in the development of models that predict how quickly rabies would spread should it ever cross the English Channel. Likewise in Sweden, where collisions involving moose (Alces alces) are among the most common traffic accidents in rural areas, research on moose behaviour has yielded ways of keeping them off roads and verges. In addition, investigations of the foraging of insect pollinators, such as honeybees, have led to impressive increases in agricultural crop yields throughout the world.

Konrad Lorenz being followed by greylag geese (Anser anser)

Even if there were no practical benefits to be gained from learning about animal behaviour, the subject would still merit exploration. Humans (Homo sapiens) are animals themselves, and most humans are deeply interested in the lives and minds of their fellow humans, their pets, and other creatures. British ethologist Jane Goodall and American field biologist George Schaller, as well as British broadcaster David Attenborough and Australian wildlife conservationist Steve Irwin, have brought the wonders of animal behaviour to the attention and appreciation of the general public. Books, television programs, and movies on the subject of animal behaviour abound.

The study of animal behavior is the scientific study of everything that animals do, whether the animals are single-celled organisms, invertebrates, fish, amphibians, reptiles, birds, humans, or other mammals. It involves the investigation of the relationship of animals to their physical environment as well as to other organisms, and includes topics such as how animals find and defend resources, avoid predators, choose mates and reproduce, and care for their young.

People who study animal behavior are typically trying to answer one or more of the following four kinds of questions about behavior (often referred to as "Tinbergen's four questions," after the animal behaviorist Niko Tinbergen who first described them):

- What is the cause of the behavior?
- What mechanism is underlying it?
- What "triggered" it just now?

- How did the behavior develop within the individual's lifetime?
- What function or functions does the behavior serve?
- How did the behavior evolve over time?

Answers to questions about the causes of behavior include both the external stimuli that affect behavior, and the internal hormonal and neural mechanisms that control behavior. Questions about the development of behavior focus on the ways in which behavior changes over the lifetime of the animal, and how these changes are affected by both genes and experience. Questions about the functions of behavior focus on both the behavior's immediate effect on the animal and on the behavior's adaptive value in helping the animal to survive or reproduce successfully in a particular environment. Finally, questions about the evolution of behavior focus on the origins of behavior patterns and how these change over generations.

Animal Sexual Behavior

The study of animal sexuality (and primate sexuality especially) is a rapidly developing field. It used to be believed that only humans and a handful of species performed sexual acts other than for procreation, and that animals' sexuality was instinctive and a simple response to the "right" stimulation (sight, scent). Current understanding is that many species that were formerly believed monogamous have now been proven to be promiscuous or opportunistic in nature; a wide range of species appear both to masturbate and to use objects as tools to help them do so; in many species animals try to give and get sexual stimulation with others where procreation is not the aim; and homosexual behaviour has now been observed among 1,500 species and in 500 of those it is well documented.

Mating Systems

In sociobiology and behavioural ecology, the term mating system is used to describe the ways in which animal societies are structured in relation to sexual behaviour. The mating system specifies which males mate with which females, and under what circumstances.

The following are some of the mating systems generally recognised in humans and other animals:

Monogamy: One male and one female have an exclusive mating relationship.

Polygamy: A single individual concurrently carries a relationship/mates with one or more of the opposite sex. Three types are recognized:

Polygyny: (the most common polygamous mating system in vertebrates so far studied): One male has an exclusive relationship with two or more females.

Polyandry: One female has an exclusive relationship with two or more males.

Polygynandry: Two or more individuals have an exclusive relationship with two or more individuals from the opposite sex; the numbers of males and females need not be equal, and in vertebrate species studied so far, there are usually fewer males.

Promiscuity: Any male and female will mate within the social group.

Monogamy: Monogamous Pairing in Animals

Zoologists and biologists now have solid evidence that monogamous pairs of animals are not always sexually exclusive. Many animals that form pairs to mate and raise offspring regularly engage in sexual activities with extra-pair partners. This includes previous exemplars such as swans. Sometimes these extra-pair sexual activities lead to offspring. Genetic tests frequently show that some of the offspring raised by a monogamous pair come from the female mating with an extra-pair male partner. These discoveries have led biologists to adopt new ways of talking about monogamy:

Social Monogamy

Social monogamy refers to a male and female's social living arrangement (e.g., shared use of a territory, behaviour indicative of a social pair, and/or proximity between a male and female) without inferring any sexual interactions or reproductive patterns. In humans, social monogamy equals monogamous marriage. Sexual monogamy is defined as an exclusive sexual relationship between a female and a male based on observations of sexual interactions. Finally, the term genetic monogamy is used when DNA analyses can confirm that a female-male pair reproduce exclusively with each other. A combination of terms indicates examples where levels of relationships coincide, e.g., sociosexual and sociogenetic monogamy describe corresponding social and sexual, and social and genetic monogamous relationships, respectively.

Whatever makes a pair of animals socially monogamous does not necessarily make them sexually or genetically monogamous. Social monogamy, sexual monogamy, and genetic monogamy can occur in different combinations.

Social monogamy is relatively rare in the animal kingdom. The actual incidence of social monogamy varies greatly across different branches of the evolutionary tree. Over 90 percent of avian species are socially monogamous.

This stands in contrast to mammals. Only 3 percent of mammalian species are socially monogamous, although up to 15 percent of primate species are socially monogamous. Social monogamy has also been observed in reptiles, fish, and insects.

Sexual monogamy is also rare among animals. Many socially monogamous species engage in extra-pair copulations, making them sexually non-monogamous. For example, while over 90% of birds are socially monogamous, "on average, 30 percent or more of the baby birds in any nest are sired by someone other than the resident male." Patricia Adair Gowaty has estimated that, out of 180 different species of socially monogamous songbirds, only 10 percent are sexually monogamous.

The incidence of genetic monogamy, determined by DNA fingerprinting, varies widely across species. For a few rare species, the incidence of genetic monogamy is 100 percent, with all offspring genetically related to the socially monogamous pair. But genetic monogamy is strikingly low in other species.

Polygamy

Polygamy is defined as a mating structure in which a single individual of one gender has exclusive access to several individuals of the opposite gender. It takes two main forms - polygyny and polyandry. As polygyny is the most common form of polygamy among vertebrates (including humans, to some extent), it has been studied far more extensively than polyandry.

In some species, notably those with harem-like structures, only one of a few males in a group of females will mate. Technically, polygyny in sociobiology and zoology is defined as a system in which a male has a relationships with more than one female, but the females are predominantly bonded to a single male.Should the active male be driven out, killed, or otherwise removed from the group, in a number of species the new male will ensure that breeding resources are not wasted on another male's young. The new male may achieve this in many different ways, including:

Competitive Infanticide

In lions and some monkeys, the new male will kill the offspring of the previous alpha male to cause their mothers to become receptive to his sexual advances since they are no longer nursing.

Harassment to Miscarriage

Amongst wild horses and baboons, the male will "systematically harass" pregnant females until they miscarry. Pheromone based spontaneous abortion. In some rodents such as mice, a new male with a different scent will cause females who are pregnant to spontaneously fail to implant recently fertilized eggs. This does not require contact; it is mediated by scent alone. It is known as the Bruce-Parkes effect.

Promiscuity

Two examples of systems in primates are promiscuous mating chimpanzees and bonobos. These species live in social groups consisting of several males and several females. Each female copulates with many males, and vice versa. In bonobos, the amount of promiscuity is particularly striking because bonobos use sex to alleviate social conflict as well as to reproduce.

Seasonal Nature of Animal Sexuality

Many animal species have specific mating (or breeding) seasons (seasonal breeding). These are often associated with changes to herd or group structure, and behavioural changes, including territorialism amongst individuals. These may be annual (e.g. wolves), biannual (e.g. dogs) or more frequently (e.g. horses). During these periods, females of most species are more mentally and physically receptive to sexual advances, a period scientifically described as estrous but commonly described as being "in season" or "in heat", but outside them animals still engage in sexual behaviours, and such acts as do occur are not necessarily harmful.

Interpretation of Animal Sexuality

The field of study of sexuality in non-human species has been a long standing taboo, with researchers either failing to observe or mis-categorizing and mis-describing sexual behaviour which does not meet their preconceptions.

More current research provides views such as that of the Natural History Museum at the University of Oslo, which in 2006 held an exhibition on animal sexuality:

Many researchers have described homosexuality as something altogether different from sex. They must realise that animals can have sex with who they will, when they will and without consideration to a researcher's ethical principles.

An example of overlooking behaviour relates to descriptions of giraffe mating:

When nine out of ten pairings occur between males, "every male that sniffed a female was reported as sex, while anal intercourse with orgasm between males was only 'revolving around' dominance, competition or greetings.

Sex for Pleasure

It is a common myth that animals do not (as a rule) have sex for pleasure, or alternatively that humans, pigs (and perhaps cats, dolphins and one or two species of primate) are the only species which do. This is sometimes formulated "animals mate only for reproduction".

Science cannot say at present conclusively what animals do or do not find "pleasurable", a question considered in more depth under Emotion in animals. The urban myth website Snopes.com considers this particular view in depth. Its conclusions are broadly that the statement is true, but only using a very specific definition of "sex for pleasure", in which sexual acts tied to a reproductive cycle or for which an alternative explanation can be asserted, are ignored, as is all sexual activity that does not involve penetration. Animals put themselves at risk to engage in sex, and as a result, most species have evolved sexual signals (usually scent and behaviour) to indicate the presence of receptive periods. During these, sex is sought, and outside these it is usually not sought (or is sought but not permitted). Snopes comments that this is not in fact a reflection of whether sex is pleasurable or not, but rather a reflection of whether individuals have sex at arbitrary times.

Types of Sexual Activity

Autoeroticism (Masturbation)

It appears that many animals, both male and female, masturbate, both when partners are available and otherwise.

For example, Petplace.com comments in its guide on assessing potential breeding stock purchases: "Masturbation is a normal behaviour in all stallions that does not reduce semen production or performance in the breeding shed"

Likewise a review from the University of Pennsylvania School of Veterinary Medicine says:

The behaviour known within the horse breeding industry as masturbation. This involves normal periodic erections and penile movements. This behavior, both from the descriptive field studies cited above and in extensive study of domestic horses, is now understood as normal, frequent behavior of male equids. Attempting to inhibit or punish masturbation, for example by tying a brush to the area of the flank underside where the penis rubs into contact with the underside, which is still a common practice of horse managers regionally around the world, often leads to increased masturbation and disturbances of normal breeding behaviour.

Castration does not prevent masturbation, as it is observed in geldings. Masturbation is common in both mares and stallions, before and after puberty.

The female porcupine has been observed to use a stick as a dildo, holding one end of a stick and walking around, straddling it as it bumped against the ground and vibrated against her genitalia.

Autoeroticism also occurs widely among animals, both male and female. A variety of creative techniques are used, including genital stimulation using the hand or front paw (primates, Lions), foot (Vampire Bats, primates), flipper (Walruses), or tail (Savanna Baboons), sometimes accompanied by stimulation of the nipples (Rhesus Macaques, Bonobos); auto-fellating or licking, sucking and/or nuzzling by a male of his own penis (Common Chimpanzees, Savanna Bonobos, Vervet Monkeys, Squirrel Monkeys, Thinhorn Sheep, Bharal, Aovdad, Dwarf Cavies); stimulation of the penis by flipping or rubbing it against the belly or in its own sheath (White-tailed and Mule Deer, Zebras and Takhi); spontaneous ejaculations (Mountain Sheep, Warthogs, Spotted Hyenas); and stimulation of the genitals using inanimate objects (found in several primates and cetaceans).

Many birds masturbate by mounting and copulating with tufts of grass, leaves or mounds of earth, and some mammals such as primates and Dolphins also rub their genitals against the ground or other surfaces to stimulate themselves.

Autoeroticism in female mammals, as well as heterosexual and homosexual intercourse (especially in primates), often involves direct or indirect stimulation of the clitoris. This organ is present in the females of all mammalian species and several other animal groups. and that:

Apes and Monkeys use a variety of objects to masturbate with and even deliberately create implements for sexual stimulation often in highly creative ways.

Petter Bøckman of the Natural History Museum at the University of Oslo commented (in respect of a 2006 exhibition on homosexuality in the animal kingdom) that: *Masturbation is common in the animal kingdom ... We have a Darwinist mentality that all animals only have sex to procreate. But there are plenty of animals who will masturbate when they have nothing better to do. Masturbation has been observed among primates, deer, killer whales and penguins, and we're talking about both males and females. They rub themselves against stones and roots. Orangutans are especially inventive. They make dildos of wood and bark.*

Oral sex

Animals of several species are documented as engaging in both autofellatio and oral sex. Although easily confused by lay-people, this is a separate and sexually oriented behaviour, distinct from non-sexual grooming or the investigation of scents.

Auto-fellatio or oral sex in animals is documented in goats, primates, hyaenas, bats and sheep.

Homosexual Behaviour - Homosexual Behavior in Animals

Mallard Ducks *Anas platyrhynchos* form male-female pairs only until the female lays eggs, at which time the male leaves the female. Mallards have rates of male-male sexual activity that are unusually high for birds, in some cases, as high as 19% of all pairs in a population. The presence of same-sex sexual behaviour was not scientifically observed on a large scale until recent times. Homosexual behaviour does occur in the animal kingdom outside humans, especially in social species, particularly in marine birds and mammals, monkeys, and the great apes. Homosexual behaviour has been observed among 1,500 species, and in 500 of those it is well documented.

To turn the approach on its head: No species has been found in which homosexual behaviour has not been shown to exist, with the exception of species that never have sex at all, such as sea urchins and aphis. Moreover, a part of the animal kingdom is hermaphroditic, truly bisexual. For them, homosexuality is not an issue.

Georgetown University professor Janet Mann has specifically theorised that homosexual behaviour, at least in dolphins, is an evolutionary advantage that minimizes intraspecies aggression, especially among males.

Male penguin couples have been documented to mate for life, build nests together, and to use a stone as a surrogate egg in nesting and brooding. In 2004, the Central Park Zoo in the United States replaced one male couple's stone with a fertilized egg, which the couple then raised as their own offspring. German and Japanese zoos have also reported homosexual behaviour among their penguins. This phenomenon has also been reported at Kelly Tarlton's Aquarium in Auckland, New Zealand. "Humans have created the myth that sexuality can be justified only by reproduction, which by definition limits it to hetero sex," says Michael Bronski, author of The Pleasure Principle: Culture, Backlash, and the Struggle for Gay Freedom. "But here is an animal society that uses homosexuality to improve its social life."

Mounting of one female by another is common among cattle.

After studying the Bonobo primates for his book Bonobo: The Forgotten Ape, primatologist Frans de Waal, a professor of psychology at Emory University in Atlanta, says that such expressions of intimacy are consistent with the homosexual behaviour of what he terms "the erotic champions of the world." "Same-sex, opposite-sex — bonobos just love sex play," de Waal said in an interview. "They have so much sex, it gets boring.

"Homosexual behaviour in male sheep (found in 6-10% of rams) is associated with variations in cerebral mass distribution and chemical activity. A study reported in Endocrinology concluded that biological and physiological factors are in effect. These findings are similar to human findings reported by Simon LeVay.

Approximately eight percent of [male] rams exhibit sexual preferences [that is, even when given a choice] for male partners (male-oriented rams) in contrast to most rams, which prefer female partners (female-oriented rams). We identified a cell group within the medial preoptic area/anterior hypothalamus of age-matched adult sheep that was significantly larger in adult rams than in ewes.

Male bighorn sheep are divisible into two kinds: the typical males among whom homosexual behaviour, including intercourse, is common and "effeminate sheep", or "behavioural transvestites", which are not known to engage in homosexual behaviour.

Cross Species Sex

While it is commonly believed that animal sexuality is instinctive and thus somewhat mechanistic, research regularly records that many animals are sexual opportunists, partaking in sexual relations with individuals of visibly distinct species. This is more visible in domesticated species and animals in captivity, as domestication commonly selects for increased breeding rate (and so an accelerated breeding cycle has commonly arisen in domesticated species over the centuries), and also because these species are more easily observed by humans. Nevertheless, animals have been observed in the wild to attempt sexual activity with other species or indeed inanimate objects. Attempts by wild moose to obtain sex from domestic horses are apparently well known by wildlife specialists.

In the wild, where observation is harder, genetic studies have shown a "large number" of inter-species hybrids, and other investigations describe productive and non-productive inter-species mating as a "natural occurrence".Recent genetic evidence strongly suggesting this has occurred even within the history of the human species, and that early humans often had sexual activity with other primate species, is considered below.

Hybrid offspring can result from two organisms of distinct but closely related parent species, although the resulting offspring is not always fertile. According to the definition of a species, if two organisms cannot or will not mate and produce a fertile offspring, they are different species. The mule, for example (a horse/donkey cross) is normally sterile, whilst the liger (lion/tiger cross) has fertile females and sterile males. Novosibirsk zoo director Rostislav Shilo says on the liger (born in its zoo); "It's just that the lion and the tiger live in neighboring caves in the Novosibirsk zoo, and got used to each other. It's practically impossible in the wild".

Due to the difficulties of observation, interspecies sex of this kind between two top-level predators, occurring in the wild, was only conclusively documented with the finding of a grizzly-polar bear hybrid in April 2006. Again, as with lions and tigers, the two species would normally not share enough common territory to provide adequate opportunity for much cross-species sexual activity.

Animal sexual advances on, and attempted interactions with, humans and other species, have been documented by ethologists such as Kohler, Gerald Durrell and Desmond Morris, as well as authoritative researchers such as Birute Galdikas who studied orangutans in Borneo. Philosopher and animal welfare activist Peter Singer reports: *While walking through the camp with Galdikas, my informant was suddenly seized by a large male orangutan, his intentions made obvious by his erect penis. Fighting off so powerful an animal was not an option, but Galdikas called to her companion not to be concerned, because the orangutan would not harm her, and adding, as further reassurance, that 'they have a very small penis' ... though the orangutan lost interest before penetration took place.*

Prostitution

In some penguin species, the females, even when in a committed relationship, will exchange sexual favours with strange males for the pebbles they need to build their nests. Prostitution was also observed among chimpanzees, who trade food for sex.

Sexual Fetishes

Although not often reported, animals, or primates at the least, are able to sexualize inanimate objects similar to the way human beings sexualize the objects of their sexual fetishes. Not only will an animal that has a habitual object for masturbation sometimes appear to sexualize that object, primates have generalized further to sexualize kinds of objects for which no instinctual or prior sexual connection exists.

Thus Gabriel, a chimpanzee at the Southwest National Primate Research Center, is said to have a shoe fetish (or possibly a leather fetish) according to caretaker Bert Barrera, and it is reported that a male chimpanzee raised in captivity developed a bit of a shoe fetish, masturbating obsessively by rubbing his caretaker's leather boot.

The sexualization of objects or locations is also well recognized in the breeding world. So for example, stallions may often 'drop' (become sexually aroused) upon visiting a location where they have been allowed to have sex before, or upon seeing a stimulus previously associated with sexual activity such as an artificial vagina.

In this case however, the primary structure is Pavlovian conditioning, and the fetishistic association is due to a conditioned response (or association) formed with a distinctive 'reward'. Human fetishism can also be traced back to similar or near-identical conditioning: likewise based upon the Pavlovian association between an erotic sensation or anticipation, and objects which become mentally associated with that activity.

Sexual Imagery Viewing

A study by Platt, Khera and Deaner at Duke University North Carolina (reported in Current Biology and online here), showed that male monkeys will give up privileges (in this case, juice, which is highly valued), to be allowed to see a female monkey's hindquarters.

Deaner and his team reported that monkeys would take a juice cut to look at powerful males' faces or the perineum of a female, but to persuade the monkeys to stare at subordinate males, the researchers had to bribe them with larger drinks. "Virtually all [male] monkeys will give up juice to see female hindquarters ... they really value the images."

The researchers stress that in monkey society, such behaviours have great social utility and we should therefore not simply reach the conclusion that "monkeys enjoy pornographic pictures". There is no evidence at this point that viewable pictures or movies of sexual activity are valued for their sexual enjoyment, although as noted above (Masturbation), there are reports that watching sex in real life may have such an effect. The subject of animals and sexual imagery is not yet well researched.

Problems with encouraging pandas to mate in captivity have been very common. However, showing young male pandas "panda pornography" is widely credited with a recent population boom among pandas in zoos.

Coercive Sex

Controversial interpretations and implications aside, sex in a forceful or apparently coercive context has also been documented in a variety of species. A notable example is bottlenose dolphins, where at times, a pod of bachelor males will 'corner' a female. Furthermore, in a zoo where it is common practice to put newly captured dolphins in with dolphins who are established in their enclosures, other species of dolphin are never put in together with bottlenoses because the bottlenose dolphins frequently torment and rape them. The behaviour is also common in some arachnids (spiders), notably those whose females eat the males during sex if not tricked with food and/or tied down with threads, and in some herbivorous herd species or species where males and females are very different in size, where the male dominates sexually by sheer force and size.

Some species of birds appear to combine sexual intercourse with apparent violent assault; these include ducks,geese, and white-fronted bee-eaters. According to Emlen and Wrege (1986) forced copulations occur in this socially nesting species, and females must avoid the unwelcome attention of males as they emerge from their nest burrows or they are forced to the ground and mated with. Apparently, such attacks are made preferentially on females who are laying and who may thus mother their offspring as a result.

In 2007, research suggested that in the Acilius genus of water beetles (also known as "diving beetles"), an "evolutionary arms race" between the genders means that there is no courtship system for these beetles. "It's a system of rape. But the females don't take things quietly. They evolve counter-weapons." Cited mating behaviours include males suffocating females underwater till exhausted, and allowing only occasional access to the surface to breathe for up to six hours (to prevent them breeding with other males), and females which have a variety of body shapings (to prevent males from gaining a grip). Foreplay is "limited to the female desperately trying to dislodge the male by swimming frantically around."

Charles Siebert reports in his New York Times article Elephant Crackup? that: Since the early 1990's, for example, young male elephants in Pilanesberg National Park and the Hluhluwe-Umfolozi Game Reserve in South Africa have been raping and killing rhinoceroses; this abnormal behaviour, according to a 2001 study in the journal Pachyderm, has been reported in "a number of reserves" in the region.

Sex between Adults and Juveniles

It has also been recorded that certain species of mole will impregnate newborns of their own species. It is not clear if this is forceful or not. Similarly, the male stoat (Mustela erminea) will mate with infant females of their species. This apparently is a natural part of their reproductive biology - there is a delayed gestation period, so these females give birth the following year when they are fully grown.

A male spotted hyena which attempted to mate with a female which succeeded in driving it off, eventually turned to its ten-month-old cub, repeatedly mounting it and ejaculating on it. The cub sometimes ignored this and sometimes struggled 'slightly as if in play'. The mother did not intervene.

Infants and children in Bonobo societies are often involved in sexual behaviour.

Sexual Cannibalism

Sexual cannibalism, which has been documented in arachnids, insects and amphipods, is a phenomenon in which a female organism kills and consumes the male before, during, or after copulation. Although it does confer some known advantages to reproduction, whether or not the male is complicit has not been scientifically determined.

Necrophilia

Necrophilia in animals is where a living animal engages in a sexual act with a dead animal. In one of the most well-known examples, Kees Moeliker of the Rotterdam Natural History Museum, Netherlands observed sexual activities outside his office between a live duck and a dead one. Two male mallards which Moeliker believed were engaged in rape flight, a common motif in duck sexual behaviour, collided with his window. "*When one died the other one just went for it and didn't get any negative feedback—well, didn't get any feedback*," according to Moeliker, who described the event as "homosexual necrophilia." The case was reported scientifically in Deinsea 8-2001, along with photos, and earned Moeliker an Ig Nobel Prize in biology, awarded for research that cannot or should not be reproduced.

Additionally, male cane toads have been documented (in Cane Toads: An Unnatural History) engaging in copulation with dead toads and inanimate objects.

Notes on specific species:

Bonobos

The Bonobo, which has a matriarchal society, is a fully bisexual species — both males and females engage in sexual behaviour with the same and the opposite sex, with females being particularly noted for engaging in sexual behaviour with each other and at up to 75% of sexual activity being nonreproductive. Primatologist Frans de Waal believes that Bonobos use sexual activity to resolve conflict between individuals. Sexual activity occurs between almost all ages and sexes of Bonobo societies.

Birds

Some black swans of Australia form sexually active male-male mated pairs and steal nests, or form temporary threesomes with females to obtain eggs, driving away the female after she lays the eggs. More of their cygnets survive to adulthood than those of different-sex pairs possibly due to their superior ability to defend large portions of land.

In early February 2004 the New York Times reported that a male pair of chinstrap penguins named Roy and Silo in the Central Park Zoo in New York City were partnered and had successfully hatched a female chick from an egg. Other penguins in New York have also been reported to be forming same-sex pairs.

Zoos in Japan and Germany have also documented male penguin couples. The couples have been shown to build nests together and use a stone to replace an egg in the nest. Researchers at Rikkyo University in Tokyo, found twenty such pairs at sixteen major aquariums and zoos in Japan. Bremerhaven Zoo in Germany attempted to break up the male couples by importing female penguins from Sweden and separating the male couples; they were unsuccessful. The zoo director stated the relationships were too strong between the couples.

Recently, a mated pair of swans in Boston were found to both be female. They too had attempted to raise eggs together. Studies have shown that ten to fifteen percent of female western gulls in some populations in the wild prefer other females.

As many as 19% of Mallard pairs in a given population have been observed to consist of male-male homosexuals.

Lizards

Whip-tailed lizard females have the ability to reproduce through parthenogenesis and as such males are rare and sexual breeding non-standard. Females engage in sexual behaviour to stimulate ovulation, with their behaviour following their hormonal cycles; during low levels of estrogen, these (female) lizards engage in "masculine" sexual roles. Those animals with currently high estrogen levels assume "feminine" sexual roles.

Lizards that perform the courtship ritual have greater fecundity than those kept in isolation due to an increase in hormones triggered by the sexual behaviours. So, even though asexual whiptail lizards populations lack males, sexual stimuli still increase reproductive success.

From an evolutionary standpoint these females are passing their full genetic code to all of their offspring rather than the 50% of genes that would be passed in sexual reproduction. Certain species of gecko also reproduce by parthenogenesis.

Flatworm

Penis fencing is a mating behaviour engaged in by certain species of flatworm, such as Pseudobiceros bedfordi. Species which engage in the practice are hermaphroditic, possessing both eggs and sperm-producing testes.

The species "fence" using two-headed dagger-like penises which are pointed, and white in color. One organism inseminates the other. The sperm is absorbed through pores in the skin, causing fertilization.

Sheep

An October 2003, study by Dr. Charles E. Roselli et al. (Oregon Health & Science University) states that homosexuality in male sheep (found in eight percent of rams) is associated with a region in the rams' brains which the authors call the "ovine Sexually Dimorphic Nucleus" (oSDN) which is half the size of the corresponding region in other male sheep.

However, some view this study to be flawed in that the determination of homosexuality within the sheep, (sample population of twenty-seven for the study), was to have animals who were unable to mount female ewes placed in a cage with two stanchioned males and two unstanchioned females (that is, the males could not move or struggle while the females could). Given the aggressive nature of the sheep copulation, the uneven treatment of males and females, many see this as simply evidence that the sheep in question were unable to be aggressive enough to mount females. Some say that the results were situational sexuality, unlike the bonds seen in human homosexuality.

The scientists found that, "The oSDN in rams that preferred females was significantly larger and contained more neurons than in male-oriented rams and ewes. In addition, the oSDN of the female-oriented rams expressed higher levels of aromatase, a substance that converts testosterone to estradiol, an estrogen hormone believed to facilitate typical male sexual behaviours. Aromatase expression was no different between male-oriented rams and ewes."

"The dense cluster of neurons that comprise the oSDN express cytochrome P450 aromatase. Aromatase mRNA levels in the oSDN were significantly greater in female-oriented rams than in ewes, whereas male-oriented rams exhibited intermediate levels of expression." These results suggest that "...naturally occurring variations in sexual partner preferences may be related to differences in brain anatomy and its capacity for estrogen synthesis." As noted previously, given the potential unagressiveness of the male population in question, the differing aromatase levels may also have been evidence of aggression levels, not sexuality. The results of this study have not been confirmed by others.

Spotted Hyena

The female Spotted Hyena has a unique urinary-genital system, closely resembling the penis of the male, called a pseudo-penis. The family structure is matriarchal and dominance relationships with strong sexual elements are routinely observed between related females.

They are notable for using visible sexual arousal as a sign of submission and not dominance, in males as well as females (females have a sizable erectile clitoris), to the extent that biologist Robert Sapolsky speculates that in order to facilitate this, their sympathetic and parasympathetic nervous systems may be partially reversed in respect to their reproductive organs.

Bottlenose Dolphins

Bottlenose Dolphin males have been observed working in pairs to follow and/or restrict the movement of a female for weeks at a time, waiting for her to become sexually receptive. The same pairs have also been observed engaging in intense sexual play with each other.

Janet Mann, a professor of biology and psychology at Georgetown University, argues that the common same-sex behaviour among male dolphin calves is about bond formation and benefits the species evolutionarily. They cite studies that have shown the dolphins later in life are bisexual and the male bonds forged from homosexuality work for protection as well as locating females with which to reproduce.

In 1991 an English man was prosecuted for allegedly having sexual contact with a dolphin. The man was found not guilty after it was revealed at trial that the dolphin was known to tow bathers through the water by hooking its large penis around them.

Seahorses

Seahorses, long upheld as monogamous and mating for life, are identified as "promiscuous, flighty, and more than a little bit gay" according to research published in 2007.

Scientists at 15 aquariums studied 90 seahorses of 3 species. Of 3168 sexual encounters, 37% were same sex acts. Flirting was common (up to 25 potential partners a day of both genders); only one species (the British Spiny Seahorse) included faithful representatives, and for these 5 of 17 were faithful, 12 were not. Bisexuality was widespread and considered "both a great surprise and a shock", with big bellied seahorses of both genders not showing partner preference. 1986 contacts were male-female, 836 were female-female and 346 were male-male.

Lions

Male lions often lead their social groups jointly with one or more of their brothers. To ensure loyalty, the male co-leaders will "strengthen the bonds" by often having sex with each other.

Horses

Anecdotal evidence suggest that some horses have environment or appearance preferences when selecting mates. There is also anecdotal evidence of limited bisexual behaviour in some stallions, although there is no conclusive scientific confirmation. The anecdotal evidence claims this is most likely to occur in a single isolated group, with no access to mares.

Humboldt Penguins

In 2009 at a zoo in Bremerhaven, Germany, two male adult humboldt penguins adopted an egg that had been abandoned by its biological parents. After the egg hatched, the two penguins raised, protected, cared for, and fed the chick in the same manner that heterosexual penguins raise their own biological offspring.

Role in Discussion of Human Sexuality

Information about animal sexuality frequently arises as a persuasive device in arguments regarding human sexuality. Originally, the lack of documented animal sexual behaviour deviant from heterosexual sexual monogamy was used to argue that the dominant heterosexual monogamy of most modern human societies is more natural and acceptable. Likewise, the lack of documented sex between animals for the purpose of pleasure was used to promote the moral standard of reserving sex primarily for procreation. Proponents of alternate sexuality attribute this early lack of documented evidence to an observer bias in researchers, who, they argue, tended to interpret sexual behaviour inconsistent with their values as other behaviour.

With increasing published evidence of different types of sexual behaviour between animals, arguments for heterosexual monogamy in human society have moved towards characterizing these behaviours as resulting from differences between humans and animals, and in particular on ambiguity in motivation and subjective experience in animals, which is difficult to study. Arguments identifying human and animal behaviour are characterized as anthropomorphism, and in some cases an opposite observer bias is attributed to researchers. Supporters of alternate sexuality embrace the new research as confirmation of the naturalness of alternate sexual behaviour and evidence of its long-term feasibility and utility.

Bunting

Bunting is a form of animal behavior, often found in cats, in which the animal butts or rubs their head against other things, including people. Bunting as a behaviour can be viewed as a variation of scent rubbing. This is when an animal, typically a carnivore, will rub its back on a scent such as prey or the urine of a conspicific. Evolutionarily speaking, scent rubbing is the oldest form of scent communication and bunting is a derivative of this behaviour. Rolling in the scent of another animal was an adaptation to camouflage the scent of a predator or outside male in order to get closer to mates. Bunting is generally considered to be a form of territorial scent-marking behaviour, where the cat rubs the scent glands on their cheeks and forehead on the object being marked. After a display of aggression, a cat will begin bunting objects in that immediate area as a form of territorial display toward a rival cat. Bunting and allorubbing are also part of feral cat behavior within colonies. An elaborate ritual which can take several minutes, two cats will rub along the side and tail of the other cat. This behaviour in domestic cats involves a system of hierarchy and may have evolved as a way to channel aggression where the costs of a conflict is too high. Cats also use bunting as a way to familiarize themselves to their environment and the pheromones secreted work to ease the cat's anxieties about an unfamiliar area.

A cat demonstrating bunting behaviour on a dog

Bunting is a normal animal behavior, and should be distinguished from head pressing, which is abnormal and typically a sign of illness.

Development

The practice of bunting stems from the behaviour that arises when kittens are very young and seek stimulation from their mother by rubbing and kneading. This behaviour develops throughout the animals life and is not only found in cats. It has also been found in other carnivorous mammals and some ungulates. Bunting in ungulates has no olfactory function but may have a role in the weaning of young. When a juvenile is nursing from its mother, it will bunt the udder with its head. This is to stimulate milk production or "let down" and causes some pain to the mother when the bunting movement is frequently performed. Overtime, there is an increase in the amount of times the udder is bunted by the young. This causes the mother to react in an attempt to prevent further pain. This reaction of the mother can be any form of defensive behaviour from a nipping bite, moving away from the young, or a jab of her horns. Bunting in domesticated and wild cat species has olfactory roots and has a range of uses which include, but are not limited to, mother-young association bonds, greeting/welcoming of kin, diffusing potential aggression in social environments, and distributing scent on areas to cultivate familiarity.

Bunting in other Species

Cattle

Bunting behaviour is a display of aggression in cattle. When two cattle are rivaling each other, they will often use bunting as a form of defense. Cattle will attempt to bunt the rival cattle with the goal of bunting their head under the hind legs of the animal. This occurs when one cattle shows submission during the final moments of a feud, and this specific behavior is calling clinching. The behaviour of bunting within cattle is first observable in calves. As a form of play fighting, a young calf will bunt the flank of its mother. A newborn calf will bunt the mother's udder and this stimulates milk flow. It has been found that when calves are taken from their mothers and raised artificially, the calf will attempt to bunt the artificial teat when milk is not being produced quick enough.

Horses

Horses will rub the bottom of their jaw onto others as a form of greeting. As seen in cattle, horse foals will bunt the mother's udder in order to stimulate milk production. Another example of bunting is when a dam experiences pain while a foal is suckling. The dam will proceed to bunt the foal with her head, non-aggressively, to prevent further discomfort. Many foals will play fight with one and other; displaying the bunting behaviour. The foal will push its head against another foal's body in an attempt to knock the other off-balance.

Lions

Lions display bunting as a form of greeting when met with other pride members

Similarly to domesticated cats, lions also use bunting as a form of greeting and territorial marking. Lions will often greet each other with this head bunting behavior when returning to a pride after a hunt. In the early stages of life, cubs procure stimulation from their mother as she cleans them by rubbing and licking them. This behaviour carries throughout their lives and bunting remains a primal source of interaction between adults as it stimulates a familiar interaction between kin.

Deimatic Behavior

Deimatic behaviour, threat display, or startle display in animals means any pattern of behaviour, such as suddenly displaying conspicuous eyespots, to scare off or momentarily distract a predator, thus giving the prey animal an opportunity to escape. The term deimatic or dymantic originates from the Greek δειματόω (deimatòo), meaning “to frighten”.

Deimatic display: Callistoctopus macropus generates a bright brownish red colour with white oval spots when disturbed

Deimatic display occurs in widely separated groups of animals, including moths, butterflies, mantises and phasmids among the insects. In the cephalopods, different species of octopuses, squids, cuttlefish and the paper nautilus are deimatic.

Displays are classified as deimatic or aposematic by the responses of the animals that see them. Where predators are initially startled but learn to eat the displaying prey, the display is classed as deimatic, and the prey is bluffing; where they continue to avoid the prey after tasting it, the display is taken as aposematic, meaning the prey is genuinely distasteful. However, these categories are not entirely mutually exclusive. It is possible for a behaviour to be both deimatic and aposematic, if it both startles a predator and indicates the presence of anti-predator adaptations.

Vertebrates including several species of frog put on warning displays; some of these species have poison glands. Among the mammals, deimatic displays are found in species with strong defences, such as in foul-smelling skunks and spiny porcupines. Such displays are often combined with warning coloration. Thus these displays in both frogs and mammals are at least in part aposematic.

In Insects

Deimatic displays are made by insects including the praying mantises (Mantidae) and stick insects (Phasmatodea). While undisturbed, these insects are usually well camouflaged. When disturbed by a potential predator, they suddenly reveal their hind wings, which are brightly coloured. In mantises, the wing display is sometimes reinforced by showing brightly coloured front legs, and accompanied by a loud hissing sound created by stridulation. For example, the grasshopper

Phymateus displays red and yellow areas on its hind wings; it is also aposematic, producing a distasteful secretion from its thorax. Similarly the threat display of the walking stick phasmid (Peruphasma schultei) is not a bluff: the insect sprays defensive dolichodial-like monoterpene chemical compounds at attackers.

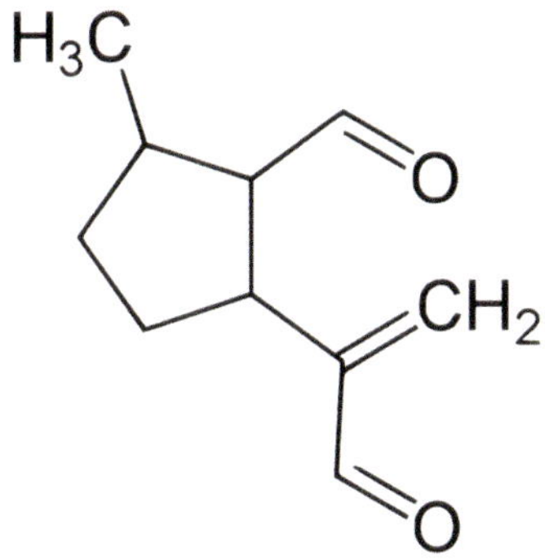

Threat displays are not always bluff. Some stick insects spray the monoterpene chemical dolichodial when attacked

A puss moth (Cerura vinula) caterpillar displaying its two flagella on its tail and red patches on its head. If the threat doesn't retreat, the caterpillar can fire formic acid from its flagella

Among moths with deimatic behaviour, the eyed hawkmoth (Smerinthus ocellatus) displays its large eyespots, moving them slowly as if it were a vertebrate predator such as an owl. Among butterflies, the peacock butterfly Aglais io is a cryptic leaf mimic with wings closed, but displays 4 conspicuous eyespots when disturbed, in a display effective against insectivorous birds (flycatchers).

An experiment by the Australian zoologist A.D. Blest demonstrated that the more an eyespot resembled a real vertebrate eye in both colour and pattern, the more effective it was in scaring off insectivorous birds. In another experiment using peacock butterflies, Blest showed that when the conspicuous eyespots had been rubbed off, insectivorous birds (yellow buntings) were much less effectively frightened off, and therefore both the sudden appearance of colour, and the actual eyespot pattern, contribute to the effectiveness of the deimatic display.

Some noctuid moths, such as the large red underwing (Catocala nupta), are cryptic at rest, but display a flash of startlingly bright colours when disturbed. Others, such as many species of genus Speiredonia and Spirama, look threatening while at rest. Also saturniid moths of the genera Attacus and Rothschildia display snake heads, but not from the frontal position.

Many arctiid moths make clicks when hunted by echolocating bats; they also often contain unpalatable chemicals. Some such as dogbane tiger moths (Cycnia tenera) have ears and conspicuous coloration, and start to make clicks when echolocating bats approach. An experiment by Canadian zoologists John M. Ratcliffe and James H. Fullard, using dogbane tiger moths and northern long-eared bats (Myotis septentrionalis), suggests the signals in fact both disrupt echolocation and warn of chemical defence. The behaviour of these insects is thus both deimatic and aposematic.

Deimatic display of the phasmid Peruphasma schultei

An adult female Mediterranean mantis, Iris oratoria, in threat pose

Speiredonia spectans resting mimicking a brooding head

Pseudocreobotra wahlbergii flashing its wings in deimatic pose

Peacock butterfly displays startling eyespots.

Threat pose of the phasmid Haaniella dehaanii

Female eyed hawk-moth, Smerinthus ocellatus, mounted to show the large eyespots

Peacock butterfly, Inachis io is a cryptic leaf mimic when its wings are closed

Spirama helicina displaying the face of a snake while at rest

A fine large "Phasma" illustrated by George Robert Gray, showing cryptic resting pose and dramatic wing flash

In Arachnids

Both spiders and scorpions are venomous, so their threat displays can be considered generally aposematic. However, some predators such as hedgehogs and spider-hunting wasps (Pompilidae) actively hunt arachnids, overcoming their defences, so when a hedgehog is startled by, for instance, the sounds made by a scorpion, there is reason to describe the display as deimatic.

Spiders make use of a variety of different threat displays. Some such as Argiope and Pholcus make themselves and their webs vibrate rapidly when they are disturbed; this blurs their outline and perhaps makes them look larger, as well as more difficult to locate precisely for an attack. Mygalomorphae spiders such as tarantulas exhibit deimatic behaviour; when threatened, the spider rears back with its front legs and pedipalps spread and fangs bared. Some species, such as the dangerous Indian ornamental tree spider (Poecilotheria regalis) have bright colouring on the front legs and mouthparts which are shown off in its threat display when it "rears up on its hind legs, and brandishes the fore limbs and palpi in the air".

Scorpions perform non-bluffing threat displays, as they have powerful defences, but various predators still eat them. When provoked, they spread their pincers and in some cases raise their abdomens, their tails standing near-erect with the sting ready for immediate use. Some scorpions in addition produce deimatic noises by stridulating with the pedipalps and first legs.

Aposematic threat display of Brazilian tarantula

Belly of the spider Poecilotheria regalis. The bright yellow forelegs are used in deimatic displays

Scorpion's threat display with pincers spread wide, abdomen raised to present sting

In Cephalopods

Deimatic behaviour is found in cephalopods including the common cuttlefish Sepia officinalis, squid such as the Caribbean reef squid (Sepioteuthis sepioidea) and bigfin reef squid (Sepioteuthis lessoniana), octopuses including the common octopus Octopus vulgaris and the Atlantic white-spotted octopus (Octopus macropus), and the paper nautilus (Argonauta argo). Deimatic cephalopod displays involve suddenly creating bold stripes, often reinforced by stretching out the animal's arms, fins or web to make it look as big and threatening as possible.

For example, in the common cuttlefish the display consists of flattening the body, making the skin pale, showing a pair of eyespots on the mantle, dark eye rings, and a dark line on the fins, and dilating the pupils of the eyes. The common octopus similarly displays pale skin and dark eye rings with dilated pupils, but also curls its arms and stretches out the web between the arms as far as possible, and squirts out jets of water. Other octopuses such as Atlantic white-spotted octopus turn bright brownish red with oval white spots all over in a high contrast display. The paper nautilus can rapidly change its appearance: it suddenly withdraws the shining iridescent web formed by its first pair of arms from its shell.

In Vertebrates

Frilled lizard faces predators, making itself look big with head frills, raising its body and waving its tail

Among vertebrates, the Australian frilled lizard (Chlamydosaurus kingii) has a startling display in which wide semicircular frills on either side of the head are fanned out; the mouth is opened wide exposing the gape; the tail is waved over the body, and the body is raised, so that the animal appears as large and threatening as possible.

Frogs such as Physalaemus nattereri, Physalaemus deimaticus, and Pleurodema brachyops have a warning display behaviour. These animals inflate themselves with air and raise their hind parts to appear as large as possible, and display brightly coloured markings and eyespots to intimidate predators. Seven species of frogs in the genus Pleurodema have lumbar glands (making the animals distasteful, so in their case the display is likely aposematic); these glands are usually boldly contrasted in black as a further warning.

Non-bluffing (aposematic) displays occur in mammals which possess powerful defences such as spines or stink glands, and which habitually warn off potential predators rather than attempting escape by running. The lowland streaked tenrec (Hemicentetes semispinosus) raises the spines on its head and back when confronted by a predator, and moves its head up and down. Porcupines such as Erethizon erect their long sharp quills and adopt a hunched, head-down posture when a predator is nearby. The spotted skunk (Spilogale putorius) balances on its front legs, its body raised vertically with its bold pelage pattern conspicuously displayed, and its tail (near the scent glands) raised and spread out.

Namaqua chameleon showing threat display with dewlap

Colombian four-eyed frog, Pleurodema brachyops

Euasian eagle owl, Bubo bubo, erects the feathers on its neck to make itself appear larger

Lowland streaked tenrec, Hemicentetes semispinosus erects spines on head and body when threatened

Striped skunk, Mephitis mephitis, displays prominent lighter markings against black, with raised bushy tail, honestly advertising its squirting scent glands

Domestic cat, Felis catus, hisses, arches its back, raises its fur to appear larger and bares its teeth in its genuine warning display

Deimatic or Aposematic?

In a study of the rattling made by rattlesnakes of different species, the Canadian zoologists Brock Fenton and Lawrence Licht found that the sounds are always similar: they have rapid onset (starting suddenly, and reaching full volume in a few milliseconds); they consist of a "broadband" mixture of frequencies between 2 kHz and 20 kHz, with little energy either in the ultrasonic (above 20 kHz) or in the rattlesnakes' hearing range (below 700 Hz); and the frequencies do not change much with time (the rattling after 2 minutes having a similar spectrum to that at onset). There was no clear difference in the sounds made by the different species measured: Crotalus horridus, Crotalus adamanteus, Crotalus atrox, Crotalus cerastes, Crotalus viridis and Sistrurus catenatus. This pattern implies that the rattling "could serve as a general attention-getting device", which "is designed as a deimatic or startle display". Its similarity to the "broadband, harsh sounds" used as warning calls by birds and mammals may enhance its effectiveness. Since rattlesnakes can barely hear the sound, it is unlikely to serve as any form of communication to other snakes of the same species. Finally, the sounds are not in themselves loud enough to cause pain and hence keep predators away.

Fenton and Licht note that the effect of a rattlesnake's rattling could be deimatic (startle) in inexperienced animals, whether predators or large animals that might injure the snake by stepping on it, but aposematic (a warning signal) in animals that are aware of the rattle's meaning. They refer to the work of Fenton and his colleague David Bates on the responses of the big brown bat, Eptesicus fuscus, to the defensive clicks made by moths in the family Arctiidae, which includes the garden tiger moth, Arctia caja. This family includes large, furry, bitter-tasting or poisonous moths. They found that while sounds can startle inexperienced bats, after a few trials the bats ignored the sounds if the prey was edible; but the same sounds can warn experienced bats of bitter-tasting prey (an honest signal).

Rattlesnake raising rattle on tail, drawn by St. George Mivart, on the Genesis of Species. The rattle may both startle inexperienced predators and warn off experienced ones

Garden tiger moth, Arctia caja, displays startling bright pattern of black spots on orange-red hindwings. The insect is bitter-tasting, so the pattern may be aposematic as well as deimatic

Collective Animal Behavior

Collective animal behavior is a form of social behavior involving the coordinated behavior of large groups of similar animals as well as emergent properties of these groups. This can include the costs and benefits of group membership, the transfer of information across the group, the group decision-making process, and group locomotion and synchronization. Studying the principles of collective animal behavior has relevance to human engineering problems through the philosophy of biomimetics. For instance, determining the rules by which an individual animal navigates relative to its neighbors in a group can lead to advances in the deployment and control of groups of swimming or flying micro-robots such as UAVs (Unmanned Aerial Vehicles).

Starling flock asunset in Denmark

Examples

Examples of collective animal behavior include:

- Flocking birds
- Herding ungulates
- Shoaling and schooling fish
- Swarming Antarctic krill
- Pods of dolphins
- Marching locusts
- Nest building ants

Proposed Functions

Many functions of animal aggregations have been proposed. These proposed functions may be grouped into the four following categories: social and genetic, anti-predator, enhanced foraging, and increased locomotion efficiency.

Social Interaction

Support for the social and genetic function of aggregations, especially those formed by fish, can be seen in several aspects of their behavior. For instance, experiments have shown that individual fish removed from a school will have a higher respiratory rate than those found in the school. This effect has been partly attributed to stress, although hydrodynamic factors were considered more important in this particular study. The calming effect of being with conspecifics may thus provide a social motivation for remaining in an aggregation. Herring, for instance, will become very agitated if they are isolated from conspecifics. Fish schools have also been proposed to serve a reproductive function since they provide increased access to potential mates.

Protection from Predators

School of goldband fusiliers

Several anti-predator functions of animal aggregations have been proposed. One potential method by which fish schools or bird flocks may thwart predators is the 'predator confusion effect' proposed and demonstrated by Milinski and Heller (1978). This theory is based on the idea that it becomes difficult for predators to pick out individual prey from groups because the many moving targets create a sensory overload of the predator's visual channel. Milinski and Heller's findings have been corroborated both in experiment and computer simulations.

A second potential anti-predator effect of animal aggregations is the "many eyes" hypothesis. This theory states that as the size of the group increases, the task of scanning the environment for predators can be spread out over many individuals. Not only does this mass collaboration presumably provide a higher level of vigilance, it could also allow more time for individual feeding.

A third hypothesis for an anti-predatory effect of animal aggregation is the "encounter dilution" effect. Hamilton, for instance, proposed that the aggregation of animals was due to a "selfish" avoidance of a predator and was thus a form of cover-seeking. Another formulation of the theory was given by Turner and Pitcher and was viewed as a combination of detection and attack probabilities. In the detection component of the theory, it was suggested that potential prey might benefit by living together since a predator is less likely to chance upon a single group than a scattered distribution. In the attack component, it was thought that an attacking predator is less likely to eat a particular animal when a greater number of individuals are present. In sum, an individual has an advantage if it is in the larger of two groups, assuming that the probability of detection and attack does not increase disproportionately with the size of the group.

Enhanced Foraging

A third proposed benefit of animal groups is that of enhanced foraging. This ability was demonstrated by Pitcher and others in their study of foraging behavior in shoaling cyprinids. In this study, the time it took for groups of minnows and goldfish to find a patch of food was quantified. The number of fishes in the groups was varied, and a statistically significant decrease in the amount of time necessary for larger groups to find food was established. Further support for an enhanced foraging capability of schools is seen in the structure of schools of predatory fish. Partridge and others analyzed the school structure of Atlantic bluefin tuna from aerial photographs and found that the school assumed a parabolic shape, a fact that was suggestive of cooperative hunting in this species.

Increased Locomotion Efficiency

This theory states that groups of animals moving in a fluid environment may save energy when swimming or flying together, much in the way that bicyclists may draft one another in a peloton. Geese flying in a Vee formation are also thought to save energy by flying in the updraft of the wing-tip vortex generated by the previous animal in the formation. Ducklings have also been shown to save energy by swimming in a line. Increased efficiencies in swimming in groups have also been proposed for schools of fish and Antarctic krill.

Group Structure

The structure of large animal groups has been difficult to study because of the large number of animals involved. The experimental approach is therefore often complemented by mathematical modeling of animal aggregations.

Experimental Approach

Experiments investigating the structure of animal aggregations seek to determine the 3D position of each animal within a volume at each point in time. It is important to know the internal structure of the group because that structure can be related to the proposed motivations for animal grouping. This capability requires the use of multiple cameras trained on the same volume in space, a technique known as stereophotogrammetry. When hundreds or thousands of animals occupy the study volume, it becomes difficult to identify each one. In addition, animals may block one another in the camera views, a problem known as occlusion. Once the location of each animal at each point in time is known, various parameters describing the animal group can be extracted.

These parameters include:

Density: The density of an animal aggregation is the number of animals divided by the volume (or area) occupied by the aggregation. Density may not be a constant throughout the group. For instance, starling flocks have been shown to maintain higher densities on the edges than in the middle of the flock, a feature that is presumably related to defense from predators.

Polarity: The group polarity describes if the group animals are all pointing in the same direction or not. In order to determine this parameter, the average orientation of all animals in the group is determined. For each animal, the angular difference between its orientation and the group orientation is then found. The group polarity is then the average of these differences.

Nearest Neighbor Distance: The nearest neighbor distance (NND) describes the distance between the centroid of one animal (the focal animal) and the centroid of the animal nearest to the focal animal. This parameter can be found for each animal in an aggregation and then averaged. Care must be taken to account for the animals located at the edge of an animal aggregation. These animals have no neighbor in one direction.

Nearest Neighbor Position: In a polar coordinate system, the nearest neighbor position describes the angle and distance of the nearest neighbor to a focal animal.

Packing Fraction: Packing fraction is a parameter borrowed from physics to define the organization (or state i.e. solid, liquid, or gas) of 3D animal groups. It is an alternative measure to density. In this parameter, the aggregation is idealized as an ensemble of solid spheres, with each animal at the center of a sphere. The packing fraction is defined as the ratio of the total volume occupied by all individual spheres divided by the global volume of the aggregation. Values range from zero to one, where a small packing fraction represents a dilute system like a gas. Cavagna found that the packing fraction for groups of starlings was 0.012.

Integrated Conditional Density: This parameter measures the density at various length scales and therefore describes the homogeneity of density throughout an animal group.

Pair Distribution Function: This parameter is usually used in physics to characterize the degree of spatial order in a system of particles. It also describes the density, but this measures describes the density at a distance away from a given point. Cavagna et al. found that flocks of starlings exhibited more structure than a gas but less than a liquid.

Modeling Approach

The simplest mathematical models of animal aggregations generally instruct the individual animals to follow three rules:

1. Move in the same direction as your neighbor
2. Remain close to your neighbors
3. Avoid collisions with your neighbors

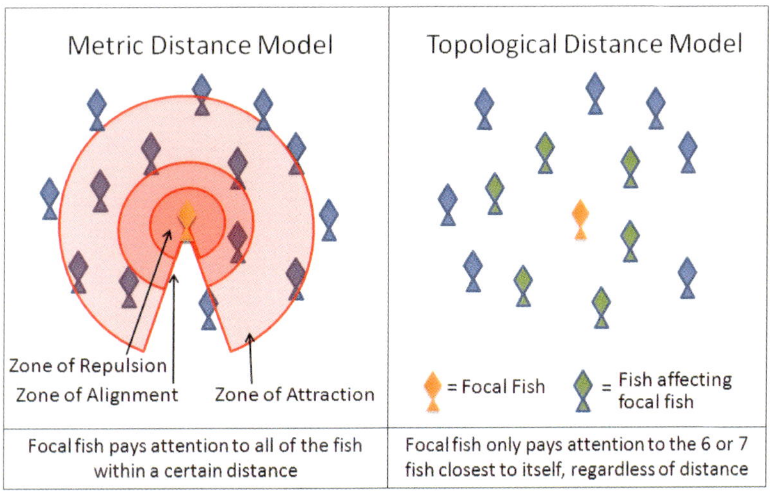

A diagram illustrating the difference between 'metric distance' and 'topological distance' in reference to fish schools

An example of such a simulation is the Boids program created by Craig Reynolds in 1986. Another is the Self Propelled Particle model. Many current models use variations on these rules. For instance, many models implement these three rules through layered zones around each animal. In the zone of repulsion very close to the animal, the focal animal will seek to distance itself from its neighbors in order to avoid a collision. In the slightly further away zone of alignment, a focal animal will seek to align its direction of motion with its neighbors. In the outmost zone of attraction, which extends as far away from the focal animal as it is able to sense, the focal animal will seeks to move towards a neighbor. The shape of these zones will necessarily be affected by the sensory capabilities of the animal. For example, the visual field of a bird does not extend behind its body. Fish, on the other hand, rely on both vision and on hydrodynamic signals relayed through its lateral line. Antarctic krill rely on vision and on hydrodynamic signals relayed through its antennae.

Recent studies of starling flocks have shown, however, that each bird modifies its position relative to the six or seven animals directly surrounding it, no matter how close or how far away those animals are. Interactions between flocking starlings are thus based on a topological rule rather than a metric rule. It remains to be seen whether the same rule can be applied to other animals. Another recent study, based on an analysis of high speed camera footage of flocks above Rome and assuming minimal behavioural rules, has convincingly simulated a number of aspects of flock behaviour.

Collective Decision Making

Aggregations of animals are faced with decisions which they must make if they are to remain together. For a school of fish, an example of a typical decision might be which direction to swim when confronted by a predator. Social insects such as ants and bees must collectively decide where to build a new nest. A herd of elephants must decide when and where to migrate. How are these decisions made? Do stronger or more experienced 'leaders' exert more influence than other group members, or does the group make a decision by consensus? The answer probably depends on the species. While the role of a leading matriarch in an elephant herd is well known, studies have shown that some animal species use a consensus approach in their collective decision-making process.

A recent investigation showed that small groups of fish used consensus decision-making when deciding which fish model to follow. The fish did this by a simple quorum rule such that individuals watched the decisions of others before making their own decisions. This technique generally resulted in the 'correct' decision but occasionally cascaded into the 'incorrect' decision. In addition, as the group size increased, the fish made more accurate decisions in following the more attractive fish model. Consensus decision-making, a form of collective intelligence, thus effectively uses information from multiple sources to generally reach the correct conclusion.

Some simulations of collective decision-making use the Condorcet method to model the way groups of animals come to consensus.

Animal Consciousness

Animal consciousness, or animal awareness, is the quality or state of self-awareness within an animal, or of being aware of an external object or something within itself. In humans, consciousness has been defined as: sentience, awareness, subjectivity, qualia, the ability to experience or to feel, wakefulness, having a sense of selfhood, and the executive control system of the mind. Despite the difficulty in definition, many philosophers believe there is a broadly shared underlying intuition about what consciousness is.

The topic of animal consciousness is beset with a number of difficulties. It poses the problem of other minds in an especially severe form because animals, lacking the ability to use human language, cannot tell us about their experiences. Also, it is difficult to reason objectively about the question, because a denial that an animal is conscious is often taken to imply that it does not feel, its life has no value, and that harming it is not morally wrong. The 17th-century French philosopher René Descartes, for example, has sometimes been blamed for mistreatment of animals because he argued that only humans are conscious.

According to the Cambridge Declaration on Consciousness, "near human-like levels of consciousness" have been observed in the grey parrot

Philosophers who consider subjective experience the essence of consciousness also generally believe, as a correlate, that the existence and nature of animal consciousness can never rigorously be known. The American philosopher Thomas Nagel spelled out this point of view in an influential essay titled *What Is it Like to Be a Bat?*. He said that an organism is conscious "if and only if there is something that it is like to be that organism—something it is like *for* the organism"; and he argued that no matter how much we know about an animal's brain and behavior, we can never really put ourselves into the mind of the animal and experience its world in the way it does itself. Other thinkers, such as the cognitive scientist Douglas Hofstadter, dismiss this argument as incoherent. Several psychologists and ethologists have argued for the existence of animal consciousness by describing a range of behaviors that appear to show animals holding beliefs about things they cannot directly perceive—Donald Griffin's 2001 book *Animal Minds* reviews a substantial portion of the evidence.

Animal consciousness has been actively researched for over one hundred years. In 1927 the American functional psychologist Harvey Carr argued that any valid measure or understanding of awareness in animals depends on "an accurate and complete knowledge of its essential conditions in man". A more recent review concluded in 1985 that "the best approach is to use experiment (especially psychophysics) and observation to trace the dawning and ontogeny of self-consciousness, perception, communication, intention, beliefs, and reflection in normal human fetuses, infants, and children". In 2012, a group of neuroscientists signed the Cambridge Declaration on Consciousness, which "unequivocally" asserted that "humans are not unique in possessing the neurological substrates that generate consciousness. Non-human animals, including all mammals and birds, and many other creatures, including octopuses, also possess these neural substrates."

Philosophical Background

René Descartes argued that only humans are conscious, and not other animals

The mind–body problem in philosophy examines the relationship between mind and matter, and in particular the relationship between consciousness and the brain. A variety of approaches have been proposed. Most are either dualist or monist. Dualism maintains a rigid distinction between the realms of mind and matter. Monism maintains that there is only one kind of stuff, and that mind and matter are both aspects of it. The problem was addressed by pre-Aristotelian philosophers, and was famously addressed by René Descartes in the 17th century, resulting in Cartesian dualism. Descartes believed that humans only, and not other animals have this non-physical mind.

The rejection of the mind–body dichotomy is found in French Structuralism, and is a position that generally characterized post-war French philosophy. The absence of an empirically identifiable meeting point between the non-physical mind and its physical extension has proven problematic to dualism and many modern philosophers of mind maintain that the mind is not something separate from the body. These approaches have been particularly influential in the sciences, particularly in the fields of sociobiology, computer science, evolutionary psychology, and the neurosciences.

Epiphenomenalism

Epiphenomenalism is the theory in philosophy of mind that mental phenomena are caused by physical processes in the brain or that both are effects of a common cause, as opposed to mental phenomena driving the physical mechanics of the brain. The impression that thoughts, feelings, or sensations cause physical effects, is therefore to be understood as illusory to some extent. For example, it is not the feeling of fear that produces an increase in heart beat, both are symptomatic of a common physiological origin, possibly in response to a legitimate external threat.

The history of epiphenomenalism goes back to the post-Cartesian attempt to solve the riddle of Cartesian dualism, i.e., of how mind and body could interact. La Mettrie, Leibniz and Spinoza all in their own way began this way of thinking. The idea that even if the animal were conscious nothing would be added to the production of behavior, even in animals of the human type, was first voiced by La Mettrie (1745), and then by Cabanis (1802), and was further explicated by Hodgson (1870) and Huxley (1874). Huxley (1874) likened mental phenomena to the whistle on a steam locomotive. However, epiphenomenalism flourished primarily as it found a niche among methodological or scientific behaviorism. In the early 1900s scientific behaviorists such as Ivan Pavlov, John B. Watson, and B. F. Skinner began the attempt to uncover laws describing the relationship between stimuli and responses, without reference to inner mental phenomena. Instead of adopting a form of eliminativism or mental fictionalism, positions that deny that inner mental phenomena exist, a behaviorist was able to adopt epiphenomenalism in order to allow for the existence of mind. However, by the 1960s, scientific behaviourism met substantial difficulties and eventually gave way to the cognitive revolution. Participants in that revolution, such as Jerry Fodor, reject epiphenomenalism and insist upon the efficacy of the mind. Fodor even speaks of "epiphobia"—fear that one is becoming an epiphenomenalist.

Thomas Henry Huxley defends in an essay titled *On the Hypothesis that Animals are Automata, and its History* an epiphenomenalist theory of consciousness according to which consciousness is a causally inert effect of neural activity—"as the steam-whistle which accompanies the work of a locomotive engine is without influence upon its machinery". To this William James objects in his essay *Are We Automata?* by stating an evolutionary argument for mind-brain interaction implying that if the preservation and development of consciousness in the biological evolution is a result of natural selection, it is plausible that consciousness has not only been influenced by neural processes, but has had a survival value itself; and it could only have had this if it had been efficacious. Karl Popper develops in the book *The Self and Its Brain* a similar evolutionary argument.

Animal Ethics

Bernard Rollin of Colorado State University, the principal author of two U.S. federal laws regulating pain relief for animals, writes that researchers remained unsure into the 1980s as to whether animals experience pain, and veterinarians trained in the U.S. before 1989 were simply taught to ignore animal pain. In his interactions with scientists and other veterinarians, Rollin was regularly asked to prove animals are conscious and provide scientifically acceptable grounds for claiming they feel pain. Academic reviews of the topic are equivocal, noting that the argument that animals have at least simple conscious thoughts and feelings has strong support, but some critics continue to question how reliably animal mental states can be determined. A refereed journal *Animal Sentience* launched in 2015 by the Institute of Science and Policy of The Humane Society of the United States is devoted to research on this and related topics.

Defining Consciousness

Consciousness is an elusive concept that presents many difficulties when attempts are made to define it. Its study has progressively become an interdisciplinary challenge for numerous researchers, including ethologists, neurologists, cognitive neuroscientists, philosophers, psychologists and psychiatrists.

> "About forty meanings attributed to the term *consciousness* can be identified and categorized based on *functions* and *experiences*. The prospects for reaching any single, agreed-upon, theory-independent definition of consciousness appear remote."

In 1976 Richard Dawkins wrote, "The evolution of the capacity to simulate seems to have culminated in subjective consciousness. Why this should have happened is, to me, the most profound mystery facing modern biology". In 2004, eight neuroscientists felt it was still too soon for a definition. They wrote an apology in "Human Brain Function":

> "We have no idea how consciousness emerges from the physical activity of the brain and we do not know whether consciousness can emerge from non-biological systems, such as computers. At this point the reader will expect to find a careful and precise definition of consciousness. You will be disappointed. Consciousness has not yet become a scientific term that can be defined in this way. Currently we all use the term *consciousness* in many different and often ambiguous ways. Precise definitions of different aspects of consciousness will emerge but to make precise definitions at this stage is premature."

Consciousness is sometimes defined as the quality or state of being aware of an external object or something within oneself. It has been defined somewhat vaguely as: subjectivity, awareness, sentience, the ability to experience or to feel, wakefulness, having a sense of selfhood, and the executive control system of the mind. Despite the difficulty in definition, many philosophers believe that there is a broadly shared underlying intuition about what consciousness is. Max Velmans and Susan Schneider wrote in *The Blackwell Companion to Consciousness*: "Anything that we are aware of at a given moment forms part of our consciousness, making conscious experience at once the most familiar and most mysterious aspect of our lives."

Related terms, also often used in vague or ambiguous ways, are:

- Awareness: the state or ability to perceive, to feel, or to be conscious of events, objects, or sensory patterns. In this level of consciousness, sense data can be confirmed by an observer without necessarily implying understanding. More broadly, it is the state or quality of being aware of something. In biological psychology, awareness is defined as a human's or an animal's perception and cognitive reaction to a condition or event.
- Self-awareness: the capacity for introspection and the ability to reconcile oneself as an individual separate from the environment and other individuals.
- Self-consciousness: an acute sense of self-awareness. It is a preoccupation with oneself, as opposed to the philosophical state of self-awareness, which is the awareness that one exists as an individual being; although some writers use both terms interchangeably or synonymously.
- Sentience: the ability to be aware (feel, perceive, or be conscious) of one's surroundings or to have subjective experiences. Sentience is a minimalistic way of defining consciousness, which is otherwise commonly used to collectively describe sentience plus other characteristics of the mind.
- Sapience: often defined as wisdom, or the ability of an organism or entity to act with appropriate judgment, a mental faculty which is a component of intelligence or alternatively may be considered an additional faculty, apart from intelligence, with its own properties.
- Qualia: individual instances of subjective, conscious experience.

Sentience (the ability to feel, perceive, or to experience subjectivity) is not the same as self-awareness (being aware of oneself as an individual). The mirror test is sometimes considered to be an operational test for self-awareness, and the handful of animals that have passed it are often considered to be self-aware. It remains debatable whether recognition of one's mirror image can be properly construed to imply full self-awareness, particularly given that robots are being constructed which appear to pass the test.

Much has been learned in neuroscience about correlations between brain activity and subjective, conscious experiences, and many suggest that neuroscience will ultimately explain consciousness; "consciousness is a biological process that will eventually be explained in terms of molecular signaling pathways used by interacting populations of nerve cells". However, this view has been criticized because consciousness has yet to be shown to be a process, and the so-called "hard problem" of relating consciousness directly to brain activity remains elusive.

Scientific Approaches

Since Descartes's proposal of dualism, it became a general consensus that the mind had become a matter of philosophy and that science was not able to penetrate the issue of consciousness - that consciousness was outside of space and time. However, over the last 20 years, many scholars have begun to move toward a science of consciousness. Antonio Damasio and Gerald Edelman are two neuroscientists who have led the move to neural correlates of the self and of consciousness. Damasio has demonstrated that emotions and their biological foundation play a critical role in high level cognition, and Edelman has created a framework for analyzing consciousness through a scientific outlook. The current problem consciousness researchers face involves explaining how and why consciousness arises from neural computation. In his research on this problem, Edelman has developed a theory of consciousness, in which he has coined the terms primary consciousness and secondary consciousness.

Eugene Linden, author of *The Parrot's Lament* suggests there are many examples of animal behavior and intelligence that surpass what people would suppose to be the boundary of animal consciousness. Linden contends that in many of these documented examples, a variety of animal species exhibit behavior that can only be attributed to emotion, and to a level of consciousness that we would normally ascribe only to our own species.

Philosopher Daniel Dennett counters that:

Consciousness requires a certain kind of informational organization that does not seem to be 'hard-wired' in humans, but is instilled by human culture. Moreover, consciousness is not a black-or-white, all-or-nothing type of phenomenon, as is often assumed. The differences between humans and other species are so great that speculations about animal consciousness seem ungrounded. Many authors simply assume that an animal like a bat has a point of view, but there seems to be little interest in exploring the details involved.

Consciousness in mammals (including humans) is an aspect of the mind generally thought to comprise qualities such as subjectivity, sentience, and the ability to perceive the relationship between oneself and one's environment. It is a subject of much research in philosophy of mind, psychology, neuroscience, and cognitive science. Some philosophers divide consciousness into phenomenal consciousness, which is subjective experience itself, and access consciousness, which refers to the global availability of information to processing systems in the brain. Phenomenal consciousness has many different experienced qualities, often referred to as qualia. Phenomenal consciousness is usually consciousness *of* something or *about* something, a property known as intentionality in philosophy of mind.

In humans, there are three common methods of studying consciousness, i.e. verbal report, behavioural demonstrations, and neural correlation with conscious activity. Unfortunately these can only be generalized to non-human taxa with varying degrees of difficulty.

Mirror Test

Elephants can recognize themselves in a mirror

The sense in which animals (or human infants) can be said to have consciousness or a self-concept has been hotly debated; it is often referred to as the debate over animal minds. The best known research technique in this area is the mirror test devised by Gordon G. Gallup, in which the skin of an animal (or human infant) is marked, while it is asleep or sedated, with a mark that cannot be seen directly but is visible in a mirror. The animal is then allowed to see its reflection in a mirror; if the animal spontaneously directs grooming behaviour towards the mark, that is taken as an indication that it is aware of itself. Over the past 30 years, many studies have found evidence that animals recognise themselves in mirrors. Self-awareness by this criterion has been reported for:

Apes

- Common chimpanzees, bonobos, orangutans and gorillas.

Other land mammals

- Elephants.

Cetaceans

- Bottlenose dolphin (*Tursiops truncatus*).
- Killer whales: Killer whales (*Orcinus orca*) and false killer whales (*Pseudorca crassidens*) may be able to recognize themselves in mirrors.

Birds

- Magpies.

Until recently it was thought that self-recognition was absent from animals without a neocortex, and was restricted to mammals with large brains and well developed social cognition. However, in 2008 a study of self-recognition in corvids reported significant results for magpies. Mammals and birds inherited the same brain components from their last common ancestor nearly 300 million years ago, and have since independently evolved and formed significantly different brain types. The results of the mirror and mark tests showed that neocortex-less magpies are capable of understanding that a mirror image belongs to their own body. The findings show that magpies respond in the mirror and mark test in a manner similar to apes, dolphins and elephants. This is a remarkable capability that, although not fully concrete in its determination of self-recognition, is at least a prerequisite of self-recognition. This is not only of interest regarding the convergent evolution of social intelligence; it is also valuable for an understanding of the general principles that govern cognitive evolution and their underlying neural mechanisms. The magpies were chosen to study based on their empathy/lifestyle, a possible precursor for their ability of self-awareness. However even in the chimpanzee, the species most studied and with the most convincing findings, clear-cut evidence of self-recognition is not obtained in all individuals tested. Occurrence is about 75% in young adults and considerably less in young and old individuals. For monkeys, non-primate mammals, and in a number of bird species, exploration of the mirror and social displays were observed. However, hints at mirror-induced self-directed behavior have been obtained.

The mirror test has attracted controversy among some researchers because it is entirely focused on vision, the primary sense in humans, while other species rely more heavily on other senses such as the olfactory sense in dogs. A study in 2015 showed that the "sniff test of self-recognition (STSR)" provides evidence of self-awareness in dogs.

Language

Another approach to determine whether a non-human animal is conscious derives from passive speech research with a macaw (see Arielle). Some researchers propose that by passively listening to an animal's voluntary speech, it is possible to learn about the thoughts of another creature and to determine that the speaker is conscious. This type of research was originally used to investigate a child's crib speech by Weir (1962) and in investigations of early speech in children by Greenfield and others (1976).

Zipf's law might be able to be used to indicate if a given dataset of animal communication indicate an intelligent natural language. Some researchers have used this algorithm to study bottlenose dolphin language.

Pain or Suffering

Further arguments revolve around the ability of animals to feel pain or suffering. Suffering implies consciousness. If animals can be shown to suffer in a way similar or identical to humans, many of the arguments against human suffering could then, presumably, be extended to animals. Others have argued that pain can be demonstrated by adverse reactions to negative stimuli that are non-purposeful or even maladaptive. One such reaction is transmarginal inhibition, a phenomenon observed in humans and some animals akin to mental breakdown.

Carl Sagan, the American cosmologist, points to reasons why humans have had a tendency to deny animals can suffer:

Humans – who enslave, castrate, experiment on, and fillet other animals – have had an understandable penchant for pretending animals do not feel pain. A sharp distinction between humans and 'animals' is essential if we are to bend them to our will, make them work for us, wear them, eat them – without any disquieting tinges of guilt or regret. It is unseemly of us, who often behave so unfeelingly toward other animals, to contend that only humans can suffer. The behavior of other animals renders such pretensions specious. They are just too much like us.

John Webster, a professor of animal husbandry at Bristol, argues:

People have assumed that intelligence is linked to the ability to suffer and that because animals have smaller brains they suffer less than humans. That is a pathetic piece of logic, sentient animals have the capacity to experience pleasure and are motivated to seek it, you only have to watch how cows and lambs both seek and enjoy pleasure when they lie with their heads raised to the sun on a perfect English summer's day. Just like humans.

However, there is no agreement where the line between organisms that can feel pain and those that cannot should be drawn. Justin Leiber, a philosophy professor at Oxford University writes that:

Montaigne is ecumenical in this respect, claiming consciousness for spiders and ants, and even writing of our duties to trees and plants. Singer and Clarke agree in denying consciousness to sponges. Singer locates the distinction somewhere between the shrimp and the oyster. He, with rather considerable convenience for one who is thundering hard accusations at others, slides by the case of insects and spiders and bacteria, they pace Montaigne, apparently and rather conveniently do not feel pain. The intrepid Midgley, on the other hand, seems willing to speculate about the subjective experience of tapeworms Nagel appears to draw the line at flounders and wasps, though more recently he speaks of the inner life of cockroaches.

There are also some who reject the argument entirely, arguing that although suffering animals feel anguish, a suffering plant also struggles to stay alive (albeit in a less visible way). In fact, no living organism 'wants' to die for another organism's sustenance. In an article written for the New York Times, Carol Kaesuk Yoon argues that:

When a plant is wounded, its body immediately kicks into protection mode. It releases a bouquet of volatile chemicals, which in some cases have been shown to induce neighboring plants to pre-emptively step up their own chemical defenses and in other cases to lure in predators of the beasts that may be causing the damage to the plants. Inside the plant, repair systems are engaged and defenses are mounted, the molecular details of which scientists are still working out, but which involve signaling molecules coursing through the body to rally the cellular troops, even the enlisting of the genome itself, which begins churning out defense related proteins ... If you think about it, though, why would we expect any organism to lie down and die for our dinner? Organisms have evolved to do everything in their power to avoid being extinguished. How long would any lineage be likely to last if its members effectively didn't care if you killed them?

Cognitive Bias and Emotion

Is the glass half empty or half full?

Cognitive bias in animals is a pattern of deviation in judgment, whereby inferences about other animals and situations may be drawn in an illogical fashion. Individuals create their own "subjective social reality" from their perception of the input. It refers to the question "Is the glass half empty or half full?", used as an indicator of optimism or pessimism. Cognitive biases have been shown in a wide range of species including rats, dogs, rhesus macaques, sheep, chicks, starlings and honeybees.

The neuroscientist Joseph LeDoux advocates avoiding terms derived from human subjective experience when discussing brain functions in animals. For example, the common practice of calling brain circuits that detect and respond to threats "fear circuits" implies that these circuits are responsible for feelings of fear. LeDoux argues that Pavlovian fear conditioning should be renamed Pavlovian threat conditioning to avoid the implication that "fear" is being acquired in rats or humans. Key to his theoretical change is the notion of survival functions mediated by survival circuits, the purpose of which is to keep organisms alive rather than to make emotions. For example, defensive survival circuits exist to detect and respond to threats. While all organisms can do this, only organisms that can be conscious of their own brain's activities can feel fear. Fear is a conscious experience and occurs the same way as any other kind of conscious experience: via cortical circuits that allow attention to certain forms of brain activity. LeDoux argues the only differences between an emotional and non-emotion state of consciousness are the underlying neural ingredients that contribute to the state.

Neuroscience

Neuroscience is the scientific study of the nervous system. It is a highly active interdisciplinary science that collaborates with many other fields. The scope of neuroscience has broadened recently to include molecular, cellular, developmental, structural, functional, evolutionary, computational, and medical aspects of the nervous system. Theoretical studies of neural networks are being complemented with techniques for imaging sensory and motor tasks in the brain. According to a 2008 paper, neuroscience explanations of psychological phenomena currently have a "seductive allure", and "seem to generate more public interest" than explanations which do not contain neuroscientific information. They found that subjects who were not neuroscience experts "judged that explanations with logically irrelevant neuroscience information were more satisfying than explanations without.

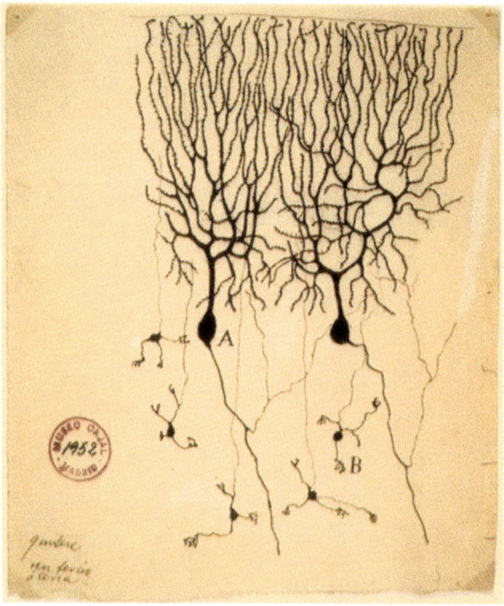

Drawing by Santiago Ramón y Cajal of neurons in the pigeon cerebellum

Neural Correlates

The neural correlates of consciousness constitute the minimal set of neuronal events and mechanisms sufficient for a specific conscious percept. Neuroscientists use empirical approaches to discover neural correlates of subjective phenomena. The set should be *minimal* because, if the brain is sufficient to give rise to any given conscious experience, the question is which of its components is necessary to produce it.

Visual sense and representation was reviewed in 1998 by Francis Crick and Christof Koch. They concluded sensory neuroscience can be used as a bottom-up approach to studying consciousness, and suggested experiments to test various hypotheses in this research stream.

A feature that distinguishes humans from most animals is that we are not born with an extensive repertoire of behavioral programs that would enable us to survive on our own ("physiological prematurity"). To compensate for this, we have an unmatched ability to learn, i.e., to consciously acquire such programs by imitation or exploration. Once consciously acquired and sufficiently exercised, these programs can become automated to the extent that their execution happens beyond the realms of our awareness. Take, as an example, the incredible fine motor skills exerted in playing a Beethoven piano sonata or the sensorimotor coordination required to ride a motorcycle along a curvy mountain road. Such complex behaviors are possible only because a sufficient number of the subprograms involved can be executed with minimal or even suspended conscious control. In fact, the conscious system may actually interfere somewhat with these automated programs.

The growing ability of neuroscientists to manipulate neurons using methods from molecular biology in combination with optical tools depends on the simultaneous development of appropriate behavioural assays and model organisms amenable to large-scale genomic analysis and manipulation. A combination of such fine-grained neuronal analysis in animals with ever more sensitive psychophysical and brain imaging techniques in humans, complemented by the development of a robust theoretical predictive framework, will hopefully lead to a rational understanding of consciousness.

Neocortex

Previously researchers had thought that patterns of neural sleep exhibited by zebra finches needed a mammalian neocortex

The neocortex is a part of the brain of mammals. It consists of the grey matter, or neuronal cell bodies and unmyelinated fibers, surrounding the deeper white matter (myelinated axons) in the cerebrum. The neocortex is smooth in rodents and other small mammals, whereas in primates and other larger mammals it has deep grooves and wrinkles. These folds increase the surface area of the neocortex considerably without taking up too much more volume. Also, neurons within the same wrinkle have more opportunity for connectivity, while neurons in different wrinkles have less opportunity for connectivity, leading to compartmentalization of the cortex. The neocortex is divided into frontal, parietal, occipital, and temporal lobes, which perform different functions. For example, the occipital lobe contains the primary visual cortex, and the temporal lobe contains the primary auditory cortex. Further subdivisions or areas of neocortex are responsible for more specific cognitive processes. The neocortex is the newest part of the cerebral cortex to evolve (hence the prefix "neo"); the other parts of the cerebral cortex are the paleocortex and archicortex, collectively known as the allocortex. In humans, 90% of the cerebral cortex is neocortex.

Researchers have argued that consciousness arises in the neocortex, and therefore cannot arise in animals which lack a neocortex. For example, Rose argued in 2002 that the "fishes have nervous systems that mediate effective escape and avoidance responses to noxious stimuli, but, these responses must occur without a concurrent, human-like awareness of pain, suffering or distress, which depend on separately evolved neocortex." Recently that view has been challenged, and many researchers now believe that animal consciousness can arise from homologous subcortical brain networks.

Attention

Attention is the cognitive process of selectively concentrating on one aspect of the environment while ignoring other things. Attention has also been referred to as the allocation of processing resources. Attention also has variations amongst cultures. Voluntary attention develops in specific cultural and institutional contexts through engagement in cultural activities with more competent community members.

Most experiments show that one neural correlate of attention is enhanced firing. If a neuron has a certain response to a stimulus when the animal is not attending to the stimulus, then when the animal does attend to the stimulus, the neuron's response will be enhanced even if the physical characteristics of the stimulus remain the same. In many cases attention produces changes in the EEG. Many animals, including humans, produce gamma waves (40–60 Hz) when focusing attention on a particular object or activity.

Extended Consciousness

Extended consciousness is an animal's autobiographical self-perception. It is thought to arise in the brains of animals which have a substantial capacity for memory and reason. It does not necessarily require language. The perception of a historic and future self arises from a stream of information from the immediate environment and from neural structures related to memory. The concept was popularised by Antonio Damasio and is used in biological psychology. Extended consciousness is said to arise in structures in the human brain described as *image spaces* and *dispositional spaces*. Image spaces imply areas where sensory impressions of all types are processed, including the focused awareness of the core consciousness. Dispositional spaces include convergence zones, which are networks in the brain where memories are processed and recalled, and where knowledge is merged with immediate experience.

Metacognition

Metacognition is defined as "cognition about cognition", or "knowing about knowing." It can take many forms; it includes knowledge about when and how to use particular strategies for learning or for problem solving. It has been suggested that metacognition in some animals provides evidence for cognitive self-awareness. There are generally two components of metacognition: knowledge about cognition, and regulation of cognition. Writings on metacognition can be traced back at least as far as *De Anima* and the *Parva Naturalia* of the Greek philosopher Aristotle. Metacognologists believe that the ability to consciously think about thinking is unique to sapient species and indeed is one of the definitions of sapience. There is evidence that rhesus monkeys and apes can make accurate judgments about the strengths of their memories of fact and monitor their own uncertainty, while attempts to demonstrate metacognition in birds have been inconclusive. A 2007 study provided some evidence for metacognition in rats, but further analysis suggested that they may have been following simple operant conditioning principles, or a behavioral economic model.

Mirror Neurons

Mirror neurons are neurons that fire both when an animal acts and when the animal observes the same action performed by another. Thus, the neuron "mirrors" the behavior of the other, as though the observer were itself acting. Such neurons have been directly observed in primate and other species including birds. The function of the mirror system is a subject of much speculation. Many researchers in cognitive neuroscience and cognitive psychology consider that this system provides the physiological mechanism for the perception action coupling. They argue that mirror neurons may be important for understanding the actions of other people, and for learning new skills by imitation. Some researchers also speculate that mirror systems may simulate observed actions, and thus contribute to theory of mind skills, while others relate mirror neurons to language abilities. Neuroscientists such as Marco Iacoboni (UCLA) have argued that mirror neuron systems in the human brain help us understand the actions and intentions of other people. In a study published in March 2005 Iacoboni and his colleagues reported that mirror neurons could discern if another person who was picking up a cup of tea planned to drink from it or clear it from the table. In addition, Iacoboni and a number of other researchers have argued that mirror neurons are the neural basis of the human capacity for emotions such as empathy. Vilayanur S. Ramachandran has speculated that mirror neurons may provide the neurological basis of self-awareness.

Evolutionary Psychology

Consciousness is likely an evolved adaptation since it meets George Williams' criteria of species universality, complexity, and functionality, and it is a trait that apparently increases fitness. Opinions are divided as to where in biological evolution consciousness emerged and about whether or not consciousness has survival value. It has been argued that consciousness emerged (i) exclusively with the first humans, (ii) exclusively with the first mammals, (iii) independently in mammals and birds, or (iv) with the first reptiles. Donald Griffin suggests in his book *Animal Minds* a gradual evolution of consciousness. Each of these scenarios raises the question of the possible survival value of consciousness.

In his paper "Evolution of consciousness," John Eccles argues that special anatomical and physical adaptations of the mammalian cerebral cortex gave rise to consciousness. In contrast, others have argued that the recursive circuitry underwriting consciousness is much more primitive, having evolved initially in pre-mammalian species because it improves the capacity for interaction with both social *and* natural environments by providing an energy-saving "neutral" gear in an otherwise energy-expensive motor output machine. Once in place, this recursive circuitry may well have provided a basis for the subsequent development of many of the functions that consciousness facilitates in higher organisms, as outlined by Bernard J. Baars. Richard Dawkins suggested that humans evolved consciousness in order to make themselves the subjects of thought. Daniel Povinelli suggests that large, tree-climbing apes evolved consciousness to take into account one's own mass when moving safely among tree branches. Consistent with this hypothesis, Gordon Gallup found that chimps and orangutans, but not little monkeys or terrestrial gorillas, demonstrated self-awareness in mirror tests.

The concept of consciousness can refer to voluntary action, awareness, or wakefulness. However, even voluntary behaviour involves unconscious mechanisms. Many cognitive processes take place in the cognitive unconscious, unavailable to conscious awareness. Some behaviours are conscious when learned but then become unconscious, seemingly automatic. Learning, especially implicitly learning a skill, can take place outside of consciousness. For example, plenty of people know how to turn right when they ride a bike, but very few can accurately explain how they actually do so.

Neural Darwinism

Neural Darwinism is a large scale theory of brain function initially proposed in 1978 by the American biologist Gerald Edelman. Edelman distinguishes between what he calls primary and secondary consciousness:

- Primary consciousness: is the ability, found in humans and some animals, to integrate observed events with memory to create an awareness of the present and immediate past of the world around them. This form of consciousness is also sometimes called "sensory consciousness". Put another way, primary consciousness is the presence of various subjective sensory contents of consciousness such as sensations, perceptions, and mental images. For example, primary consciousness includes a person's experience of the blueness of the ocean, a bird's song, and the feeling of pain. Thus, primary consciousness refers to being mentally aware of things in the world in the present without any sense of past and future; it is composed of mental images bound to a time around the measurable present.
- Secondary consciousness: is an individual's accessibility to their history and plans. The concept is also loosely and commonly associated with having awareness of one's own consciousness. The ability allows its possessors to go beyond the limits of the remembered present of primary consciousness.

Primary consciousness can be defined as simple awareness that includes perception and emotion. As such, it is ascribed to most animals. By contrast, secondary consciousness depends on and includes such features as self-reflective awareness, abstract thinking, volition and metacognition.

Edelman's theory focuses on two nervous system organizations: the brainstem and limbic systems on one side and the thalamus and cerebral cortex on the other side. The brain stem and limbic system take care of essential body functioning and survival, while the thalamocortical system receives signals from sensory receptors and sends out signals to voluntary muscles such as those of the arms and legs. The theory asserts that the connection of these two systems during evolution helped animals learn adaptive behaviors.

Other scientists have argued against Edelman's theory, instead suggesting that primary consciousness might have emerged with the basic vegetative systems of the brain. That is, the evolutionary origin might have come from sensations and primal emotions arising from sensors and receptors, both internal and surface, signaling that the well-being of the creature was immediately threatened—for example, hunger for air, thirst, hunger, pain, and extreme temperature change. This is based on neurological data showing the thalamic, hippocampal, orbitofrontal, insula, and midbrain sites are the key to consciousness of thirst. These scientists also point out that the cortex might not be as important to primary consciousness as some neuroscientists have believed. Evidence of this lies in the fact that studies show that systematically disabling parts of the cortex in animals does not remove consciousness. Another study found that children born without a cortex are conscious. Instead of cortical mechanisms, these scientists emphasize brainstem mechanisms as essential to consciousness. Still, these scientists concede that higher order consciousness does involve the cortex and complex communication between different areas of the brain.

While animals with primary consciousness have long-term memory, they lack explicit narrative, and, at best, can only deal with the immediate scene in the remembered present. While they still have an advantage over animals lacking such ability, evolution has brought forth a growing complexity in consciousness, particularly in mammals. Animals with this complexity are said to have secondary consciousness. Secondary consciousness is seen in animals with semantic capabilities, such as the four great apes. It is present in its richest form in the human species, which is unique in possessing complex language made up of syntax and semantics. In considering how the neural mechanisms underlying primary consciousness arose and were maintained during evolution, it is proposed that at some time around the divergence of reptiles into mammals and then into birds, the embryological development of large numbers of new reciprocal connections allowed rich re-entrant activity to take place between the more posterior brain systems carrying out perceptual categorization and the more frontally located systems responsible for value-category memory. The ability of an animal to relate a present complex scene to its own previous history of learning conferred an adaptive evolutionary advantage. At much later evolutionary epochs, further re-entrant circuits appeared that linked semantic and linguistic performance to categorical and conceptual memory systems. This development enabled the emergence of secondary consciousness.

Ursula Voss of the Universität Bonn believes that the theory of protoconsciousness may serve as adequate explanation for self-recognition found in birds, as they would develop secondary consciousness during REM sleep. She added that many types of birds have very sophisticated language systems. Don Kuiken of the University of Alberta finds such research interesting as well as if we continue to study consciousness with animal models (with differing types of consciousness), we would be able to separate the different forms of reflectiveness found in today's world.

For the advocates of the idea of a secondary consciousness, self-recognition serves as a critical component and a key defining measure. What is most interesting then, is the evolutionary appeal that arises with the concept of self-recognition. In non-human species and in children, the mirror test has been used as an indicator of self-awareness.

Cambridge Declaration on Consciousness

In 2012, a group of neuroscientists attending a conference on "Consciousness in Human and non-Human Animals" at the University of Cambridge in the UK, signed *The Cambridge Declaration on Consciousness.*

"The absence of a neocortex does not appear to preclude an organism from experiencing affective states. Convergent evidence indicates that non-human animals have the neuroanatomical, neurochemical, and neurophysiological substrates of conscious states along with the capacity to exhibit intentional behaviors. Consequently, the weight of evidence indicates that humans are not unique in possessing the neurological substrates that generate consciousness. Non-human animals, including all mammals and birds, and many other creatures, including octopuses, also possess these neurological substrates."

In the accompanying text they "unequivocally" asserted:

- "The field of Consciousness research is rapidly evolving. Abundant new techniques and strategies for human and non-human animal research have been developed. Consequently, more data is becoming readily available, and this calls for a periodic reevaluation of previously held preconceptions in this field. Studies of non-human animals have shown that homologous brain circuits correlated with conscious experience and perception can be selectively facilitated and disrupted to assess whether they are in fact necessary for those experiences. Moreover, in humans, new non-invasive techniques are readily available to survey the correlates of consciousness."
- "The neural substrates of emotions do not appear to be confined to cortical structures. In

fact, subcortical neural networks aroused during affective states in humans are also critically important for generating emotional behaviors in animals. Artificial arousal of the same brain regions generates corresponding behavior and feeling states in both humans and non-human animals. Wherever in the brain one evokes instinctual emotional behaviors in non-human animals, many of the ensuing behaviors are consistent with experienced feeling states, including those internal states that are rewarding and punishing. Deep brain stimulation of these systems in humans can also generate similar affective states. Systems associated with affect are concentrated in subcortical regions where neural homologies abound. Young human and non-human animals without neocortices retain these brain-mind functions. Furthermore, neural circuits supporting behavioral/electrophysiological states of attentiveness, sleep and decision making appear to have arisen in evolution as early as the invertebrate radiation, being evident in insects and cephalopod mollusks (e.g., octopus)."

- "Birds appear to offer, in their behavior, neurophysiology, and neuroanatomy a striking case of parallel evolution of consciousness. Evidence of near human-like levels of consciousness has been most dramatically observed in grey parrots. Mammalian and avian emotional networks and cognitive microcircuitries appear to be far more homologous than previously thought. Moreover, certain species of birds have been found to exhibit neural sleep patterns similar to those of mammals, including REM sleep and, as was demonstrated in zebra finches, neurophysiological patterns previously thought to require a mammalian neocortex. Magpies in particular have been shown to exhibit striking similarities to humans, great apes, dolphins, and elephants in studies of mirror self-recognition."
- "In humans, the effect of certain hallucinogens appears to be associated with a disruption in cortical feedforward and feedback processing. Pharmacological interventions in non-human animals with compounds known to affect conscious behavior in humans can lead to similar perturbations in behavior in non-human animals. In humans, there is evidence to suggest that awareness is correlated with cortical activity, which does not exclude possible contributions by subcortical or early cortical processing, as in visual awareness. Evidence that human and non-human animal emotional feelings arise from homologous subcortical brain networks provide compelling evidence for evolutionarily shared primal affective qualia."

Examples

A common image is the *scala naturae*, the ladder of nature on which animals of different species occupy successively higher rungs, with humans typically at the top. A more useful approach has been to recognize that different animals may have different kinds of cognitive processes, which are better understood in terms of the ways in which they are cognitively adapted to their different ecological niches, than by positing any kind of hierarchy.

Mammals

Dogs

Dogs were previously listed as non-self-aware animals. Traditionally, self-consciousness was evaluated via the mirror test. But dogs, as many other animals, are not visually oriented. In 2015, one study showed that the "sniff test of self-recognition" (STSR) provides significant evidences of self-awareness in dogs, and could play a crucial role in showing that this capacity is not a specific feature of only great apes, humans and a few other animals, but it depends on the way in which researchers try to verify it. According to the biologist Roberto Cazzolla Gatti, "the innovative approach to test the self-awareness with a smell test highlights the need to shift the paradigm of the anthropocentric idea of consciousness to a species-specific perspective". This study hasn't been replicated as of yet.

Birds

Grey Parrots

Research with captive grey parrots, especially Irene Pepperberg's work with an individual named Alex, has demonstrated they possess the ability to associate simple human words with meanings, and to intelligently apply the abstract concepts of shape, colour, number, zero-sense, etc. According to Pepperberg and other scientists, they perform many cognitive tasks at the level of dolphins, chimpanzees, and even human toddlers. Another notable African grey is N'kisi, which in 2004 was said to have a vocabulary of over 950 words which she used in creative ways. For example, when Jane Goodall visited N'kisi in his New York home, he greeted her with "Got a chimp?" because he had seen pictures of her with chimpanzees in Africa.

In 2011, research led by Dalila Bovet of Paris West University Nanterre La Défense, demonstrated grey parrots were able to coordinate and collaborate with each other to an extent. They were able to solve problems such as two birds having to pull strings at the same time to obtain food. In another example, one bird stood on a perch to release a food-laden tray, while the other pulled the tray out from the test apparatus. Both would then feed. The birds were observed waiting for their partners to perform the necessary actions so their behaviour could be synchronized. The parrots appeared to express individual preferences as to which of the other test birds they would work with.

Corvids

The Eurasian magpie passes the mirror test

It was recently thought that self-recognition was restricted to mammals with large brains and highly evolved social cognition, but absent from animals without a neocortex. However, in 2008, an investigation of self-recognition in corvids was conducted revealing the ability of self-recognition in the magpie. Mammals and birds inherited the same brain components from their last common ancestor nearly 300 million years ago, and have since independently evolved and formed significantly different brain types. The results of the mirror test showed that although magpies do not have a neocortex, they are capable of understanding that a mirror image belongs to their own body. The findings show that magpies respond in the mirror test in a manner similar to apes, dolphins, killer whales, pigs and elephants. This is a remarkable capability that, although not fully concrete in its determination of self-recognition, is at least a prerequisite of self-recognition. This is not only of interest regarding the convergent evolution of social intelligence, it is also valuable for an understanding of the general principles that govern cognitive evolution and their underlying neural mechanisms. The magpies were chosen to study based on their empathy/lifestyle, a possible precursor for their ability of self-awareness.

Invertebrates

Octopus travelling with shells collected for protection

Octopuses are highly intelligent, possibly more so than any other order of invertebrates. The level of their intelligence and learning capability are debated, but maze and problem-solving studies show they have both short- and long-term memory. Octopus have a highly complex nervous system, only part of which is localized in their brain. Two-thirds of an octopus' neurons are found in the nerve cords of their arms. Octopus arms show a variety of complex reflex actions that persist even when they have no input from the brain. Unlike vertebrates, the complex motor skills of octopuses are not organized in their brain using an internal somatotopic map of their body, instead using a non-somatotopic system unique to large-brained invertebrates. Some octopuses, such as the mimic octopus, move their arms in ways that emulate the shape and movements of other sea creatures.

In laboratory studies, octopuses can easily be trained to distinguish between different shapes and patterns. They reportedly use observational learning, although the validity of these findings is contested. Octopuses have also been observed to play: repeatedly releasing bottles or toys into a circular current in their aquariums and then catching them. Octopuses often escape from their aquarium and sometimes enter others. They have boarded fishing boats and opened holds to eat crabs. At least four specimens of the veined octopus (*Amphioctopus marginatus*) have been witnessed retrieving discarded coconut shells, manipulating them, and then reassembling them to use as shelter.

Animal Communication

Animal communication, process by which one animal provides information that other animals can incorporate into their decision making. The vehicle for the provision of this information is called a signal. The signal may be a sound, colour pattern, posture, movement, electrical discharge, touch, release of an odorant, or some combination of these mediums.

Animals communicate by sending and receiving signals. For example, a mother dingo (Canis lupus dingo) can communicate certain types of information to her pups by using tactile signals conveyed through grooming

Animals face daily decisions about how to behave. Choices can be as simple as a sea anemone deciding when to expand its tentacles or as complex as a male lion deciding whether to approach a reluctant mate. The decision, which may be reflexive or conscious, is guided by evolutionary biases based on alternative outcomes of choice, recent experience about likely conditions, and sensory information. An animal with access to complete information can always choose correctly. However, life is rarely so accommodating, and inputs often fail to provide complete information. Thus, communication is an important source of additional information that is incorporated into the decision-making process.

Cactus wrens (Campylorhynchus brunneicapillus) sing to communicate with other members of their species

Signals are actions or anatomical structures whose primary function is the provision of information to another animal. However, not all actions by one animal that provide information to another animal qualify as signals. The noise created by a foraging mouse and used by an owl to locate and kill the mouse is a case in point. Mice have to feed, and the noises they create while feeding (e.g., through movement and chewing) are an inadvertent result of that activity. Thus, these sounds are not a signal. In contrast, the song of a wren is not inadvertent—wrens sing solely to communicate with other birds.

Senders and Receivers

An animal that provides a signal is called a sender. The animal to which the signal is directed is the receiver. The receiver uses the signal information to help make a decision. For example, if a receiver must choose either to fight with or to flee from an opponent, it brings to this decision biases and thresholds passed on to it by successful prior generations. This information helps the receiver avoid harm and find food, shelter, and mates. Prior experience in the receiver's own life may also play a role in shaping its evaluation of the situation. If it has routinely lost fights to larger animals, a useful strategy would be to assess the size of the opponent. This may be done by using vision or other means. For example, in some cases an opponent broadcasts a low-frequency sound signal at the receiver. Because only large animals can produce low-frequency sounds, this signal provides evidence that the opponent is large. The receiver integrates its perception of the sound frequency with its prior experience and inherited avoidance of harmful situations and thus decides to flee.

In this example, the receiver can interpret the signal only if it understands that low-frequency sounds tend to be associated with large body sizes. The association between alternative signals (e.g., sounds of different frequencies) and different alternative circumstances (e.g., relative sizes of opponents) is called a code. Codes can be characterized as probabilities that a sender will emit a given signal in any given circumstance. In a perfect code, only one signal will be used in a given context, and only one context will evoke that signal. Real codes do not need to be perfect, but they do need to be good enough that a receiver attending to signals makes better decisions than if it ignored the signals and relied only on other sources of information.

Animals differ widely in the mechanisms by which they acquire signal codes. Some codes are inherited genetically. For example, the sound-producing structures of many male insects generate a limited range of sound frequencies, and the ears of females are pretuned to be most sensitive to those frequencies. In other species, senders' sounds or body odours are determined by random genetic processes, and receivers must learn which signals go with which individuals. Many songbirds have genetic limits on the range of sounds they can sing, but they can learn one or more local variants within those limits during a short period in their youth. In certain species, such as parrots or humans, both sender and receiver must learn the appropriate vocal coding, and they can continue to learn alternative coding systems throughout life.

Different contexts require different kinds of information and thus different signals. The number of signals in a species' repertoire can range from 5 or 6 in the simplest nonsocial animals to 10–20 in social insects, such as bees and ants, or to 30–40 in social vertebrates, such as wolves and primates. Most animals produce signals to attract mates and then produce additional signals to synchronize mating. Signals for mediating conflicts, including signals of aggressive intention and signals of submission, are also widespread. In addition, territorial species require signals for declaring territory ownership, and in situations in which adults guard or feed their young, both parents and offspring require signals to coordinate parental care. Social animals may use signals to coordinate group movements, to assemble dispersed group members, or to display social affiliations. Some animals have special signals that they use to share food finds, to alert others about predator attacks, and even to alert approaching predators that they have been detected. In addition, bats, oilbirds, porpoises, and electric fish use the differences between their own emitted and subsequently received signals to extract information about the ambient environment. In many of these contexts, the relevant animal signals are designed to provide a receiver with ancillary information about the identity, sex, social affiliation, and location of the sender.

Wolves are social animals, and they thus require a large repertoire of signals to communicate different kinds of information

Modes

A lamb investigates a rabbit, an example of interspecific communication using body posture and olfaction

Visual

- Gestures: The best known form of communication involves the display of distinctive body parts, or distinctive bodily movements; often these occur in combination, so a movement acts to reveal or emphasize a body part. A notable example is the presentation of a parent herring gull's bill to its chick as a signal for feeding. Like many gulls, the herring gull has a brightly coloured bill, yellow with a red spot on the lower mandible near the tip. When the parent returns to the nest with food, it stands over its chick and taps the bill on the ground; this elicits a begging response from a hungry chick (pecking at the red spot), which stimulates the parent to regurgitate food. The complete signal therefore involves a distinctive morphological feature (body part), the red-spotted bill, and a distinctive movement (tapping towards the ground) which makes the red spot highly visible to the chick. While all primates use some form of gesture, Frans de Waal concluded that apes and humans are unique in that only they use intentional gestures to communicate. He tested the hypothesis that gestures evolve into language by studying the gestures of bonobos and chimps.

A dog's visual tooth-baring threat display, accompanied by an auditory signal, growling

- Facial expression: Facial gestures play an important role in animal communication. Often a facial gesture is a signal of emotion. Dogs, for example, express anger through snarling and showing their teeth. In alarm their ears perk up, in fear the ears flatten while the dogs expose their teeth slightly and squint their eyes. Jeffrey Mogil studied the facial expressions of mice during increments of increasing pain; there were five recognizable facial expressions; orbital tightening, nose and cheek bulge, and changes in ear and whisker carriage.
- Gaze following: Social animals coordinate their communication by monitoring of each other's head and eye orientation. Such behaviour has long been recognized as an important component of communication during human development, and gaze-following has recently received much attention in animals. Studies have been conducted on apes, monkeys, dogs, birds, and tortoises, and have focused on two different tasks: "follow[ing] another's gaze into distant space" and "follow[ing] another's gaze geometrically around a visual barrier e.g. by repositioning themselves to follow a gaze cue when faced with a barrier blocking their view". The first ability has been found among a broad range of animals, while the second has been demonstrated only for apes, dogs, wolves, and corvids (ravens); attempts to demonstrate this "geometric gaze following" in marmoset and ibis gave negative results. Researchers do not yet have a clear picture of the cognitive basis of gaze following, but developmental evidence indicates that "simple" gaze following and "geometric" gaze following probably rely on different cognitive mechanisms.
- Color change: Color change can be separated into changes that occur during growth and development, and those triggered by mood, social context, or abiotic factors such as temperature. The latter are seen in many taxa. Some cephalopods, such as the octopus and the cuttlefish, have specialized skin cells (chromatophores) that can change the apparent colour, opacity, and reflectiveness of their skin. In addition to their use for camouflage, rapid changes in skin color are used while hunting and in courtship rituals. Cuttlefish may display two entirely different signals simultaneously from opposite sides of their body.

When a male cuttlefish courts a female in the presence of other males, he displays a male pattern facing the female and a female pattern facing away, to deceive other males. Some color signals occur in cycles. For example, when a female olive baboon begins to ovulate, her anogenital area swells and turns a bright red/pink. This signals to males that she is ready to mate.

- Bioluminescent communication: Communication by the production of light occurs commonly in vertebrates and invertebrates in the oceans, particularly at depths (e.g. angler fish). Two well known forms of land bioluminescence occur in fireflies and glow worms. Other insects, insect larvae, annelids, arachnids and even species of fungi possess bioluminescent abilities. Some bioluminescent animals produce the light themselves whereas others have a symbiotic relationship with bioluminescent bacteria.

Auditory

Bird calls can serve as alarms or keep members of a flock in contact, while the longer and more complex bird songs are associated with courtship and mating

Many animals communicate through vocalization. Vocal communication serves many purposes, including mating rituals, warning calls, conveying location of food sources, and social learning. In a number of species, males perform calls during mating rituals as a form of competition against other males and to signal females. Examples include hammer-headed bats, red deer, humpback whales, elephant seals, and songbirds. Other instances of vocal communication include the alarm calls of the Campbell monkey, the territorial calls of gibbons, and the use of frequency in greater spear-nosed bats to distinguish between groups. The vervet monkey gives a distinct alarm call for each of its four different predators, and the reactions of other monkeys vary appropriately according to the call. For example, if an alarm call signals a python, the monkeys climb into the trees, whereas the "eagle" alarm causes monkeys to seek a hiding place on the ground. Prairie dogs also use complex calls that signal predator differences. According to Con Slobodchikoff and others, prairie dog calls communicate the type, size, and speed of an approaching predator. Whale vocalizations have been found to have different dialects based on region.

Not all animals use vocalization as a means of auditory communication. Many arthropods rub specialized body parts together to produce sound. This is known as stridulation. Crickets and grasshoppers are well known for this, but many others use stridulation as well, including crustaceans, spiders, scorpions, wasps, ants, beetles, butterflies, moths, millipedes, and centipedes. Another means of auditory communication is the vibration of swim bladders in bony fish. The structure of swim bladders and the attached sonic muscles varies greatly across bony fish families, resulting in a wide variety of sounds. Striking body parts together can also produce auditory signals. A well-known example of this is the tail tip vibration of rattlesnakes as a warning signal. Other examples include bill clacking in birds, wing clapping in manakin courtship displays, and chest beating in gorillas.

Olfactory

Despite being the oldest method of communication, chemical communication is one of the least understood forms due in part to the sheer abundance of chemicals in our environment and the difficulty of detecting and measuring all the chemicals in a sample. The ability to detect chemicals in the environment serves many functions, a crucial one being the detection of food, a function that first arose in single-celled organisms (bacteria) living in the oceans during the early days of life on Earth. As this function evolved, organisms began to differentiate between chemicals compounds emanating from resources, conspecifics (same species; i.e., mates and kin), and heterospecifics (different species; i.e., competitors and predators). For instance, a small minnow species may do well to avoid habitat with a detectable concentration of chemical cue associated with a predator species such as northern pike. Minnows with the ability to perceive the presence of predators before they are close enough to be seen and then respond with adaptive behaviour (such as hiding) are more likely to survive and reproduce. Scent marking and scent rubbing are common forms of olfactory communication in mammals.

Electric

Electrocommunication is a rare form of communication in animals. It is seen primarily in aquatic animals, though some land mammals, notably the platypus and echidnas, sense electric fields that might be used for communication.

Weakly electric fishes provide an example of electrocommunication, together with electrolocation. These fish use an electric organ to generate an electric field, which is detected by electroreceptors. Differences in the waveform and frequency of changes in the field convey information on species, sex, and identity. These electric signals can be generated in response to hormones, circadian rhythms, and interactions with other fish. Some predators, such as sharks and rays, are able to eavesdrop on these electrogenic fish through passive electroreception.

Touch

Touch is a key factor in many social interactions. Here are some examples:

- Fighting: In a fight, touch may be used to challenge an opponent and to coordinate movements during the fight. It may also be used by the loser to indicate submission.
- Mating: Mammals often initiate mating by grooming, stroking or rubbing against each other. This provides the opportunity to apply chemical signals and to assess those excreted by the potential mate. Touch may also announce the intention of the male to mount the female, as when a male kangaroo grabs the tail of a female. During mating, touch stimuli are important for pair positioning, coordination and genital stimulation.

- Social integration: Touch is widely used for social integration, a use that is typified by the social grooming of one animal by another. Social grooming has several functions; it removes parasites and debris from the groomed animal, it reaffirms the social bond or hierarchical relationship between the animals, and it gives the groomer an opportunity to examine olfactory cues on the groomed individual, perhaps adding additional ones. This behaviour has been observed in social insects, birds and mammals.
- Foraging: Some ant species recruit fellow workers to new food finds by first tapping them with their antennae and forelegs, then leading them to the food source while keeping physical contact. Another example of this is the waggle dance of honey bees.
- Huddling: Prolonged physical contact or huddling also serves social integration. Huddling promotes heat exchange, together with the transfer of olfactory or tactile information. Some organisms live in constant contact in a colony, for example colonial corals. When individuals are linked tightly in this way an entire colony can react on the aversive or alarm movements made by only a few individuals. In several herbivorous insect nymphs and larvae, aggregations where there is prolonged contact play a major role in group coordination. These aggregations may take the form of a procession or a rosette.

Seismic

Seismic communication is the exchange of information using self-generated vibrational signals transmitted via a substrate such as the soil, water, spider webs, plant stems, or a blade of grass. This form of communication has several advantages, for example it can be sent regardless of light and noise levels, and it usually has a short range and short persistence, which may reduce the danger of detection by predators. The use of seismic communication is found in many taxa, including frogs, kangaroo rats, mole rats, bees, nematode worms, and others. Tetrapods usually make seismic waves by drumming on the ground with a body part, a signal that is sensed by the sacculus of the receiver. The sacculus is an organ in the inner ear containing a membranous sac that is used for balance, but can also detect seismic waves in animals that use this form of communication. Vibrations may be combined with other sorts of communication.

Thermal

A number of different snakes have the ability to sense infrared (IR) thermal radiation, which allows these reptiles to derive thermal images from the radiant heat emitted by predators or prey at wavelengths between 5 and 30 μm. The accuracy of this sense is such that a blind rattlesnake can target its strike to the vulnerable body parts of a prey animal. It was previously thought that the pit organs evolved primarily as prey detectors, but it is now believed that they may also be used to control body temperature.

The facial pits enabling thermoregulation underwent parallel evolution in pitvipers and some boas and pythons, having evolved once in pitvipers and multiple times in boas and pythons. The electrophysiology of the structure is similar between lineages, but it differs in gross structure anatomy. Most superficially, pitvipers possess one large pit organ on either side of the head, between the eye and the nostril (loreal pit), while boas and pythons have three or more comparatively smaller pits lining the upper and sometimes the lower lip, in or between the scales. Those of the pitvipers are the more advanced, having a suspended sensory membrane as opposed to a simple pit structure. Within the family Viperidae, the pit organ is seen only in the subfamily Crotalinae: the pitvipers. Despite the detection of IR radiation, the pits' IR mechanism is dissimilar to photoreceptors; while photoreceptors detect light via photochemical reactions, the protein in the facial pits of snakes is a temperature sensitive ion channel. It senses infrared signals through a mechanism involving warming of the pit organ, rather than chemical reaction to light. This is consistent with the thin pit membrane, which allows incoming IR radiation to quickly and precisely warm a given ion channel and trigger a nerve impulse, as well as vascularize the pit membrane to rapidly cool the ion channel back to its original "resting" or "inactive" temperature.

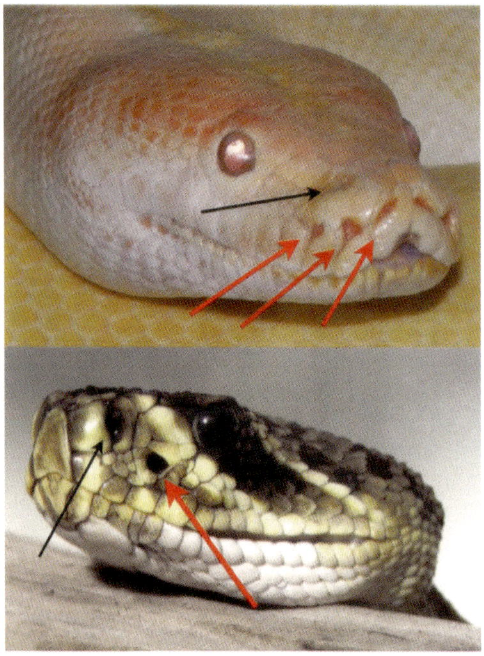

A python (top) and rattlesnake illustrating the positions of the pit organs. Red arrows indicate the pit organs whereas black arrows indicate the nostril

Common vampire bats (*Desmodus rotundus*) have specialized IR sensors in their nose-leaf. Vampire bats are the only mammals that feed exclusively on blood. The IR sense enables Desmodus to localize homeothermic animals such as cattle and horses within a range of about 10 to 15 cm. This infrared perception may be used in detecting regions of maximal blood flow on targeted prey.

Bee Learning and Communication

Honey bees are sensitive to odors (including pheromones), tastes, and colors, including ultraviolet. They can demonstrate capabilities such as color discrimination through classical and operant conditioning and retain this information for several days at least; they communicate the location and nature of sources of food; they adjust their foraging to the times at which food is available; they may even form cognitive maps of their surroundings.

Bees learn and communicate in a variety of ways

Learning

Learning is essential for efficient foraging. Honey bees are unlikely to make many repeat visits if a plant provides little in the way of reward. A single forager will visit different flowers in the morning and, if there is sufficient attraction and reward in a particular kind of flower, she will make visits to that type of flower for most of the day, unless the plants stop producing reward or weather conditions change. Honey bees are quite adept at associative learning, and many of the standard phenomena of classical conditioning take the same form in honey bees as they do in the vertebrates that are the more usual subjects of such experiments.

Swarming bees require good communication to all congregate in the same place

Foragers were trained to enter a simple Y-shaped maze that had been marked at the entrance with a particular color. Inside the maze was a branching point where the bee was required to choose between two paths. One path, which led to the food reward, was marked with the same color that had been used at the entrance to the maze, while the other was marked with a different color. Foragers learned to choose the correct path, and continued to do so when a different kind of marker (black and white stripes oriented in various directions) was substituted for the colored markers. When the experimental conditions were reversed, rewarding bees for choosing the inner passage marked with a symbol that was different from the entrance symbol, the bees again learned to choose the correct path. Extending the length of the tunnel to increase the time between seeing the one marker indicating the correct path and a second marker identifying the correct path show that the bees can retain the information in their visual working memory for about 5 seconds, equivalent to the short-term memory of birds.

Color Learning in Honeybees

One of the most common ways that honey bees, *Apis mellifera*, demonstrate associative learning is in the context of color recognition and discrimination tasks. Just as vertebrate species such as mice or pigeons that can be trained to perform associative learning tasks, honey bees make excellent subjects for tasks involving discrimination and color memory. Beginning in the early 1900s, scientists Karl von Frisch and later Randolf Menzel began asking questions about the existence, learning rates, memory, and timing of color vision in bees.

Color Discrimination

Bee Vision vs. Human Vision

The Austrian zoologist Karl von Frisch began the exploration of color vision in honey bees when he asked the first question in 1919: does color vision in bees exist? By making use of bees associative learning abilities he performed an elegant experiment to show that honey bees were in fact capable of color discrimination.

To test color vision, von Frisch first trained his honey bees to feed from a small dish filled with a nectar-like sugar water. This dish was placed over a small piece of blue colored cardboard so that the color was visible to the bees as they fed. Once the bees had become accustomed to the blue cardboard, von Frisch surrounded the blue piece of cardboard with other identically sized pieces in varying shades of gray and placed small dishes over each piece. If bees could not discriminate between colors, they would be unable to distinguish the blue piece from the many gray toned pieces. In the case that bees did not have color vision then, von Frisch predicted that the bees would visit the gray and blue pieces with equal frequency, as they would not be able to tell a difference between them.

When he allowed bees access to the dishes, however, he found that the vast majority of the bees flew directly to the blue piece of cardboard on which they had previously obtained their sugar-water reward. The bees largely ignored the gray pieces which had not been rewarded. This directed exploration and targeting of the blue cardboard showed the bees could indeed discriminate between the gray and blue shades, showing that bees do possess color vision. Von Frisch repeated this same basic experiment to show that bees produced the same results with other colors like violet and yellow. Later other researchers were able to apply this excellent experimental design to other vertebrates as well, making it an invaluable insight into testing color vision in many organisms.

Color Learning Rates and Preferences

After von Frisch's initial studies, the German scientist Randolf Menzel continued the study of color vision in honey bees and performed more detailed tests. He was curious about which colors honey bees would be able to learn fastest and whether or not bees had a greater aptitude for learning certain colors.

He used lights of varying color and intensity to illuminate circles of light on a solid surface. This set up was similar to the pieces of colored cardboard employed by von Frisch, but by using light instead of cardboard, Menzel was able to change the intensity and color of light easily. He could simply adjust the projection of the light to create a wide variety of different experimental set-ups.

Honey bee collecting pollen

To test the intricacies of the bee color vision von Frisch had established, Menzel performed an experiment that aimed to test bees ability to distinguish between two different colors. To do this, Menzel used a projected circle of colored light surrounding a small dish that could hold a sugar-water reward. Menzel then projected a second circle of differently colored light surrounding a second dish some distance away from the first. Next, a single bee was placed equidistant between these two different lights and allowed to choose which dish to search for a sugar-water treat. Only one of the colored light circles surrounded a dish that contained sugar-water; the other was empty. Menzel was then able to measure how quickly the bees learned to preferentially search only the rewarded light and to ignore the dish surrounded by unrewarded light.

The results of the experiment showed that bees did not learn to discriminate between all color pairs equally well. The fastest rate of learning was when violet light was rewarded. The color that the bees had the most difficulty learning was green, and all other colors fell somewhere in between. This evidence of inherent bias is evolutionarily reasonable given that color vision in bees allows them to distinguish between different nectar-bearing flowers, much like the rewarded dishes. As more flowers are purple than green it makes sense that bees would be more sensitive to colors likely to result in nourishment.

Color Memory

After this work on color preferences, Menzel extended his experiments to study memory in honey bee color vision. He wanted to know how many trials were necessary for honey bees to reliably choose a previously rewarded color when presented with several choices for potential rewards and how long honey bees could retain information about which color would be rewarded.

To test these questions, Menzel performed a variety of experiments. First, he presented individual bees with a sugar reward on a colored background for just a single trial. He then kept these bees in small cages for several days without any further trials. After a few days, he presented each bee with several dishes each on a different colored background at once. One of the colors was the same as that used during the initial trial. The others were novel, unrewarded colors. Amazingly, after just one trial and several days without any exposure to the rewarded color, bees correctly chose to explore the color used in the first trial more than fifty-percent of the time.

Menzel then repeated this experiment with another group of bees, keeping all factors the same except that in the second round of testing he gave the bees three initial trials with the rewarded color instead of just one. After several days in confinement when the bees were presented with a choice of colors just as in the first experiment, they virtually always chose the color that had been used during the first three trials.

This ability to retain information about color-linked rewards over a period of several days and after only minimal exposure to the colored background indicates the great strength of honeybees memory with respect to color vision.

Timing in Color Learning

One of von Frisch's students, Elizabeth Opfinger, observed that bees would learn color when approaching a feeder. Menzel took this question further: when do bees register and learn color? He wanted to know if bees registered color before, during, or after receiving their sugar-water reward. In order to explore this intriguing question, Menzel displayed the color beneath a rewarded dish at different stages of the honey bee feeding process: during approach, feeding and departure.

The outcome of this experiment revealed that bees register color during both the approach and feeding stages of the exposure process. In order for a bee to accurately remember a given color, it must be present for approximately five seconds in total. Although it varies slightly, Menzel and his colleagues found that bees usually remember best when the stimulus is present for about three seconds during the approach and two seconds after landing and beginning to feed.

Neurobiology of Color Vision

Color vision in honey bees can also be examined from a neurobiological perspective in terms of the structure and organization of their compound eyes.

In 1975 Menzel published a seminal paper describing the morphology and spectral sensitivity of the honey bee eye. He examined color-coding the honey bee retina by using a technique to mark individual cells with a fluorescent dye and record from these cells as single units. Such fine structure analysis allowed him to determine that there are three types of receptors in the honey bee eye:

1) UV receptors, 2) blue receptors, and 3) green receptors. The three receptors are dominated by three rhodopsin-like pigments. These pigments have maximal absorbance at wavelengths corresponding to 350 nm, 440 nm, and 540 nm.

Western honey bee

As the cells were examined in detail, certain features were distinguishable for each type of receptor cell. UV cells were found to form the longest visual fibres. These long visual fibers penetrated the lamina with arborizations, a tree-like branching of the fibers and spines. Blue and green receptor cells have more shallow fibers.

Interestingly, Menzel found that most of the cells he studied had secondary sensitivities that corresponded to wavelength regions at which the other two receptor types were maximally active. He used spectral efficiency experiments to show that such corresponding wavelength receptivity is the result of electric coupling.

Communication

Foragers communicate their floral findings in order to recruit other worker bees of the hive to forage in the same area. The factors that determine recruiting success are not completely known but probably include evaluations of the quality of nectar and/or pollen brought in.

There are two main hypotheses to explain how foragers recruit other workers—the "waggle dance" or "dance language" theory and the "odor plume" theory. The dance language theory is far more widely accepted, and has far more empirical support. The theories also differ in that the former allows for an important role of odor in recruitment (i.e., effective recruitment relies on dance *plus* odor), while the latter claims that the dance is essentially irrelevant (recruitment relies on odor alone). The academic debate between these two theories is extremely polarized and often hostile.

Dance Communication

It has long been known that successfully foraging Western honey bees perform a dance on their return to the hive, known as *waggle dance*, indicating that food is farther away, while the *round dance* is a short version of the *waggle dance*, indicating that food is nearby. The laden forager dances on the comb in a circular pattern, occasionally crossing the circle in a zig-zag or waggle pattern. Aristotle described this behaviour in his *Historia Animalium*. It was thought to attract the attention of other bees.

Eight-Shaped waggle dance of the honeybee (Apis mellifera). A waggle run oriented 45° to the right of 'up' on the vertical comb indicates a food source 45° to the right of the direction of the sun outside the hive. The abdomen of the dancer appears blurred because of the rapid motion from side to side

In 1947, Karl von Frisch correlated the runs and turns of the dance to the distance and direction of the food source from the hive. The orientation of the dance correlates to the relative position of the sun to the food source, and the length of the waggle portion of the run is correlated to the distance from the hive. Also, the more vigorous the display is, the better the food. There is no evidence that this form of communication depends on individual learning. Honeybees detect the dances of conspecifics by sensing near field sound and electric fields using the Johnston's organ.

Von Frisch performed a series of experiments to validate his theory. He was awarded the Nobel Prize in Physiology or Medicine in 1973 for his discoveries.

One of the most important lines of evidence on the origin and utility of the dance is that all of the known species and races of honey bees exhibit the behavior, but details of its execution vary among the different species. For example, in *Apis florea* and *Apis andreniformis* (the "dwarf honeybees") the dance is performed on the dorsal, horizontal portion of the nest, which is exposed. The runs and dances point directly toward the resource in these species. Each honey bee species has a characteristically different correlation of "waggling" to distance, as well. Such species-specific behavior suggests that this form of communication does not depend on learning but is rather determined genetically. It also suggests how the dance may have evolved.

Various experiments document that changes in the conditions under which the dance is performed lead to characteristic changes in recruitment to external resources, in a manner consistent with von Frisch's original conclusions. Experiments with robotic dummies were indeed to induce some recruitment, which should not have been possible if the dance contains no information. Various experimental results demonstrate that the dance does convey information, but the use of this information may be context-dependent, and this may explain why the results of earlier studies were inconsistent. Researchers have also discovered other forms of honeybee dance communication, such as the tremble dance.

Odor Plume

While many researchers believe that bee dances give enough information to locate resources, proponents of the odor plume theory argue that the dance gives little, or no actual guidance to a nectar source. They argue that bees instead are primarily recruited by odor. The purpose of the dance is simply to gain attention to the returning worker bee so she can share the odor of the nectar with other workers who will then follow the odor trail to the source. Most scientists agree that odor is used in recruitment to resources, but they differ strongly in opinion as to the information content of the dance.

The primary lines of evidence used by the odor plume advocates are

1. experiments with odorless sugar sources which show that worker bees are unable to recruit to those sources and
2. logical difficulties of a small-scale dance (a few centimeters across) giving directions precise enough to hold the other bees on course during a flight that could be several kilometers long. Misreading by even a few degrees would lead the bee off course by hundreds of meters at the far end.

Neither of these points invalidate the dance theory, but simply suggest that odor might be involved, which is indeed conceded by all proponents of dance theory. Critics of the odor plume theory counter that most natural nectar sources are relatively large—orchards or entire fields— so, precision may not be necessary or even desirable. They have also challenged the reproducibility of the odorless source experiment.

Odor learning in bees is usually tested by the proboscis extension reflex. Significant to the argument are the experiments of William F. Towne, of the Kutztown University in Pennsylvania, in which hives are moved to "mirror image" terrain settings, and the bees are thereby fooled into dancing about the wrong location for a nectar source. Foragers were successfully recruited to the wrong location, but only when the sun was obscured by clouds, forcing them to rely on terrain-based navigation rather than "solar ephemeris"-based navigation. As the cloud cover broke up, more and more bees corrected their dances to indicate the actual location of nectar, and forager visits shifted to the correct location.

Odor is an essential and even necessary at various stages of the recruitment process, including once a recruited forager reaches the vicinity of the resource while some scientists think that dancing may be a simple idiothetic movement that conveys no information. Others see the dance as conveying information, but doing it poorly compared to other means and potentially used backup approach.

Note: much of the research on the two competing hypotheses of communication has been restricted to Western honey bees. Other species of *Apis* use variants on the same theme, and other types of bees use other methods altogether.

Trophallaxis

The exchange of food, trophallaxis, can be used to communicate the quality of a food source, temperature, a need for water, and the condition of the queen.

Primer Pheromones

Research that was published in November 2004, by scientists under the leadership of Zachary Huang, Michigan State University indicates that so called primer pheromones play an important part in how a honey bee colony adjusts its distribution of labor most beneficially. In order to survive as a bee colony of sometimes 50,000–100,000 individual bees, the communal structure has to be adaptable to seasonal changes and the availability of food. The division of labor has to adjust itself to the resources available from foraging. While the division of labor in a bee colony is quite complex, the work can be roughly seen as work inside the hive and outside the hive. Younger bees play a role inside the hive while older bees play a role outside the hive mostly as foragers. Huang's team found that forager bees gather and carry a chemical called ethyl oleate in the stomach. The forager bees feed this primer pheromone to the worker bees, and the chemical keeps them in a nurse bee state. The pheromone prevents the nurse bees from maturing too early to become forager bees. As forager bees die off, less of the ethyl oleate is available and nurse bees more quickly mature to become foragers. It appears that this control system is an example of decentralized decision making in the bee colony.

Other bees like *Trigona corvina* rely on pheromones for much of their communication with nest mates and rivals. They produce pheromones from their labial glands. The function of signaling depends on the profitability, but they commonly will scent mark a food source either for self-orientation, to deter rivals or to direct a nest mate to the resource. Once an individual finds a good food source, they will return to the same source for many days. If an individual detects the scent of a rival bee, they will avoid the plant in order to avoid conflict and to save time. It has also been shown that pheromones are a method of sexual selection between male drones and queens.

Cognition

Experiments by James Gould suggest that honey bees may have a cognitive map for information they have learned, and utilize it when communicating.

In one test reported in a 1983 issue of *Science News*, he moved a supply of sugar water 25% further away from a hive each day. The bees communicated to each other as usual on its location. Then he placed the sugar water on a boat anchored in the middle of a small lake. When scouts returned to the hive to communicate their find, other bees refused to go with them, even though they frequently flew over the lake to reach pollen sources on the opposite shore.

In another test related in the August 1986 issue of *Discover* ("A Honey of a Question: Are Bees Intelligent?"), Gould lured some bees to a dish of artificial nectar, then gradually moved it farther from the hive after they became accustomed to it. He marked the trained bees, placed them in a darkened jar, and relocated them to a spot where the hive was still visible, but not the dish. When released one by one, the bees would appear disoriented for a few seconds, then fly directly for the covert dish. Seventy-three of 75 bees reached it in about 28 seconds. They apparently accomplished this feat by devising a new flight path based on a cognitive map of visible landmarks.

Comparative Psychology

Comparative psychology refers to the scientific study of the behavior and mental processes of non-human animals, especially as these relate to the phylogenetic history, adaptive significance, and development of behavior. Research in this area addresses many different issues, uses many different methods and explores the behavior of many different species from insects to primates.

Comparative psychology is sometimes assumed to emphasize cross-species comparisons, including those between humans and animals. However, some researchers feel that direct comparisons should not be the sole focus of comparative psychology and that intense focus on a single organism to understand its behavior is just as desirable; if not more so. Donald Dewsbury reviewed the works of several psychologists and their definitions and concluded that the object of comparative psychology is to establish principles of generality focusing on both proximate and ultimate causation.

Using a comparative approach to behavior allows one to evaluate the target behavior from four different, complementary perspectives, developed by Niko Tinbergen. First, one may ask how pervasive the behavior is across species (i.e. how common is the behavior between animal species?). Second, one may ask how the behavior contributes to the lifetime reproductive success of the individuals demonstrating the behavior (i.e. does the behavior result in animals producing more offspring than animals not displaying the behavior)? Theories addressing the ultimate causes of behavior are based on the answers to these two questions.

Third, what mechanisms are involved in the behavior (i.e. what physiological, behavioral, and environmental components are necessary and sufficient for the generation of the behavior)? Fourth, a researcher may ask about the development of the behavior within an individual (i.e. what maturational, learning, social experiences must an individual undergo in order to demonstrate a behavior)? Theories addressing the proximate causes of behavior are based on answers to these two questions.

Species Studied

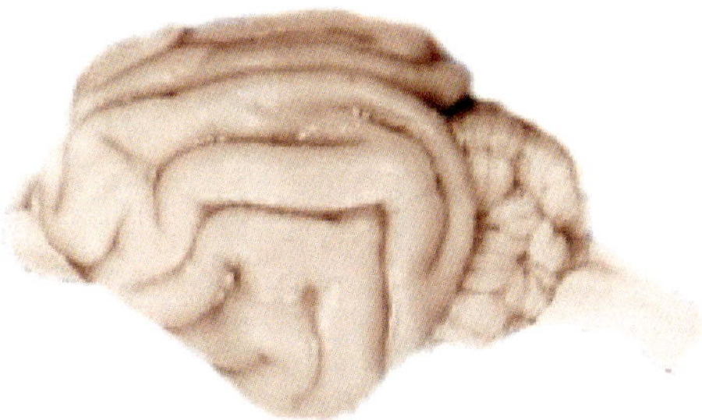

the brain of a cat

A wide variety of species have been studied by comparative psychologists. However, a small number have dominated the scene. Ivan Pavlov's early work used dogs; although they have been the subject of occasional studies, since then they have not figured prominently. Increasing interest in the study of abnormal animal behavior has led to a return to the study of most kinds of domestic animal. Thorndike began his studies with cats, but American comparative psychologists quickly shifted to the more economical rat, which remained the almost invariable subject for the first half of the 20th century and continues to be used.

Skinner introduced the use of pigeons, and they continue to be important in some fields. There has always been interest in studying various species of primate; important contributions to social and developmental psychology were made by Harry F. Harlow's studies of maternal deprivation in rhesus monkeys. Cross-fostering studies have shown similarities between human infants and infant chimpanzees. Kellogg and Kellogg (1933) aimed to look at heredity and environmental effects of young primates. They found that a cross-fostered chimpanzee named Gua was better at recognizing human smells and clothing and that the Kelloggs' infant (Donald) recognised humans better by their faces. The study ended 9 months after it had begun, after the infant began to imitate the noises of Gua.

Nonhuman primates have also been used to show the development of language in comparison with human development. For example, Gardner (1967) successfully taught the female chimpanzee Washoe 350 words in American Sign Language. Washoe subsequently passed on some of this teaching to her adopted offspring, Loulis. A criticism of Washoe's acquisition of sign language focused on the extent to which she actually understood what she was signing. Her signs may have just based on an association to get a reward, such as food or a toy. Other studies concluded that apes do not understand linguistic input, but may form an intended meaning of what is being communicated. All great apes have been reported to have the capacity of allospecific symbolic production.

Interest in primate studies has increased with the rise in studies of animal cognition. Other animals thought to be intelligent have also been increasingly studied. Examples include various species of corvid, parrots — especially the grey parrot — and dolphins. Alex (Avian Learning EXperiment) is a well known case study (1976–2007) which was developed by Pepperberg, who found that the African gray parrot Alex did not only mimic vocalisations but understood the concepts of same and different between objects. The study of non-human mammals has also included the study of dogs. Due to their domestic nature and personalities, dogs have lived closely with humans, and parallels in communication and cognitive behaviours have therefore been recognised and further researched. Joly-Mascheroni and colleagues (2008) demonstrated that dogs may be able to catch human yawns and suggested a level of empathy in dogs, a point that is strongly debated. Pilley and Reid found that a Border Collie named Chaser was able to successfully identify and retrieve 1022 distinct objects/toys.

Animal Cognition

Researchers who study animal cognition are interested in understanding the mental processes that control complex behavior, and much of their work parallels that of cognitive psychologists working with humans. For example, there is extensive research with animals on attention, categorization, concept formation, memory, spatial cognition, and time estimation. Much research in these and other areas is related directly or indirectly to behaviors important to survival in natural settings, such as navigation, tool use, and numerical competence. Thus, comparative psychology and animal cognition are heavily overlapping research categories.

Disorders of Animal Behavior

Veterinary surgeons recognize that the psychological state of a captive or domesticated animal must be taken into account if its behavior and health are to be understood and optimized.

Common causes of disordered behavior in captive or pet animals are lack of stimulation, inappropriate stimulation, or overstimulation. These conditions can lead to disorders, unpredictable and unwanted behavior, and sometimes even physical symptoms and diseases. For example, rats who are exposed to loud music for a long period will ultimately develop unwanted behaviors that have been compared with human psychosis, like biting their owners.

The way dogs behave when understimulated is widely believed to depend on the breed as well as on the individual animal's character. For example, huskies have been known to ruin gardens and houses if they are not allowed enough activity. Dogs are also prone to psychological damage if they are subjected to violence. If they are treated very badly, they may become dangerous.

The systematic study of disordered animal behavior draws on research in comparative psychology, including the early work on conditioning and instrumental learning, but also on ethological studies of natural behavior. However, at least in the case of familiar domestic animals, it also draws on the accumulated experience of those who have worked closely with the animals.

Human-animal Relationships

The relationship between humans and animals has long been of interest to anthropologists as one pathway to an understanding the evolution of human behavior. Similarities between the behavior of humans and animals have sometimes been used in an attempt to understand the evolutionary significance of particular behaviors. Differences in the treatment of animals have been said to reflect a society's understanding of human nature and the place of humans and animals in the scheme of things. Domestication has been of particular interest. For example, it has been argued that, as animals became domesticated, humans treated them as property and began to see them as inferior or fundamentally different from humans.

Ingold remarks that in all societies children have to learn to differentiate and separate themselves from others. In this process, strangers may be seen as "not people," and like animals. Ingold quoted Sigmund Freud: "Children show no trace of arrogance which urges adult civilized men to draw a hard-and-fast line between their own nature and that of all other animals. Children have no scruples over allowing animals to rank as their full equals." With maturity however, humans find it hard to accept that they themselves are animals, so they categorize, separating humans from animals, and animals into wild animals and tame animals, and tame animals into house pets and livestock. Such divisions can be seen as similar to categories of humans: who is part of a human community and someone who isn't, that is, the outsider.

The New York Times ran an article that showed the psychological benefits of animals, more specifically of children with their pets. It's been proven that having a pet does in fact improve kids' social skills. In the article, Dr. Sue Doescher, a psychologist involved in the study, stated, "It made the children more cooperative and sharing." It was also shown that these kids were more confident with themselves and able to be more empathic with other children.

Furthermore, in an edition of *Social Science and Medicine* it was stated, "A random survey of 339 residents from Perth, Western Australia were selected from three suburbs and interviewed by telephone. Pet ownership was found to be positively associated with some forms of social contact and interaction, and with perceptions of neighborhood friendliness. After adjustment for demographic variables, pet owners scored higher on social capital and civic engagement scales." Results like these let us know that owning a pet provides opportunities for neighborly interaction, among many other chances for socialization among people.

Emotion in Animals

Charles Darwin was one of the first scientists to write about the existence and nature of emotions in animals. His observational (and sometimes anecdotal) approach has developed into a more robust, hypothesis-driven, scientific approach. General hypotheses relating to correlates between humans and animals also support the claim that animals may feel emotions and that human emotions evolved from the same mechanisms. Several tests, such as cognitive bias tests and learned helplessness models, have been developed. Cognitive biases (feelings of optimism or pessimism) have been shown in a wide range of species including rats, dogs, cats, rhesus macaques, sheep, chicks, starlings, pigs and honeybees.

A drawing of a cat by T. W. Wood in Charles Darwin's book *The Expression of the Emotions in Man and Animals*, described as acting "in an affectionate frame of mind"

Some behaviourists, such as John B. Watson, claim that stimulus–response models provide a sufficient explanation for animal behaviours that have been described as emotional, and that all behaviour, no matter how complex, can be reduced to a simple stimulus-response association. Watson described that the purpose of psychology was "to predict, given the stimulus, what reaction will take place; or given the reaction, state what the situation or stimulus is that has caused the reaction".

Definitions, and Differentiation

The word "emotion" dates back to 1579, when it was adapted from the French word *émouvoir*, which means "to stir up". However, the earliest precursors of the word likely date back to the very origins of language.

Emotions have been described as discrete and consistent responses to internal or external events which have a particular significance for the organism. Emotions are brief in duration and consist of a coordinated set of responses, which may include physiological, behavioural, and neural mechanisms. Emotions have also been described as the result of evolution because they provided good solutions to ancient and recurring problems that faced ancestors.

Laterality

It has been proposed that negative, withdrawal-associated emotions are processed predominantly by the right hemisphere, whereas the left hemisphere is largely responsible for processing positive, approach-related emotions. This has been called the "laterality-valence hypothesis".

Basic and Complex Human Emotions

In humans, a distinction is sometimes made between "basic" and "complex" emotions. Six emotions have been classified as basic: anger, disgust, fear, happiness, sadness and surprise. Complex emotions would include contempt, jealousy and sympathy. However, this distinction is difficult to maintain, and animals are often said to express even the complex emotions.

Background

Behaviourist Approach

A squirrel communicating with its pup

Prior to the development of animal sciences such as comparative psychology and ethology, interpretation of animal behaviour tended to favour a minimalistic approach known as behaviourism. This approach refuses to ascribe to an animal a capability beyond the least demanding that would explain a behaviour; anything more than this is seen as unwarranted anthropomorphism. The behaviourist argument is, why should humans postulate consciousness and all its near-human implications in animals to explain some behaviour, if mere stimulus-response is a sufficient explanation to produce the same effects?

The cautious wording of Dixon exemplifies this viewpoint:

> " Recent work in the area of ethics and animals suggests that it is philosophically legitimate to ascribe emotions to animals. Furthermore, it is sometimes argued that emotionality is a morally relevant psychological state shared by humans and non humans. What is missing from the philosophical literature that makes reference to emotions in animals is an attempt to clarify and defend some particular account of the nature of emotion, and the role that emotions play in a characterization of human nature. I argue in this paper that some analyses of emotion are more credible than others. Because this is so, the thesis that humans and nonhumans share emotions may well be a more difficult case to make than has been recognized thus far. "

Moussaieff Masson and McCarthy describe a similar view (with which they disagree):

> "While the study of emotion is a respectable field, those who work in it are usually academic psychologists who confine their studies to human emotions. The standard reference work, *The Oxford Companion to Animal Behaviour*, advises animal behaviourists that "One is well advised to study the behaviour, rather than attempting to get at any underlying emotion. There is considerable uncertainty and difficulty related to the interpretation and ambiguity of emotion: an animal may make certain movements and sounds, and show certain brain and chemical signals when its body is damaged in a particular way. But does this mean an animal feels—is *aware* of—pain as we are, or does it merely mean it is programmed to act a certain way with certain stimuli? Similar questions can be asked of any activity an animal (including a human) might undertake, in principle. Many scientists regard all emotion and cognition (in humans and animals) as having a purely mechanistic basis."

Because of the philosophical questions of consciousness and mind that are involved, many scientists have stayed away from examining animal and human emotion, and have instead studied measurable brain functions through neuroscience.

Comparative Approach

In 1903, C. Lloyd Morgan published Morgan's Canon, a specialised form of Occam's razor used in ethology, in which he stated:

> "In no case is an animal activity to be interpreted in terms of higher psychological processes, if it can be fairly interpreted in terms of processes which stand lower in the scale of psychological evolution and development."

Darwin's Approach

Charles Darwin initially planned to include a chapter on emotion in *The Descent of Man* but as his ideas progressed they expanded into a book, *The Expression of the Emotions in Man and Animals*. Darwin proposed that emotions are adaptive and serve a communicative and motivational function, and he stated three principles that are useful in understanding emotional expression: First, *The Principle of Serviceable Habits* takes a Lamarckian stance by suggesting that emotional expressions that are useful will be passed on to the offspring. Second, *The Principle of Antithesis* suggests that some expressions exist merely because they oppose an expression that is useful. Third, *The Principle of the Direct Action of the Excited Nervous System on the Body* suggests that emotional expression occurs when nervous energy has passed a threshold and needs to be released.

Darwin saw emotional expression as an outward communication of an inner state, and the form of that expression often carries beyond its original adaptive use. For example, Darwin remarks that humans often present their canine teeth when sneering in rage, and he suggests that this means that a human ancestor probably utilized their teeth in aggressive action. A domestic dog's simple tail wag may be used in subtly different ways to convey many meanings as illustrated in Darwin's *The Expression of the Emotions in Man and Animals* published in 1872.

Examples of tail position indicating different emotions in dogs

"Small dog watching a cat on a table"

"Dog approaching another dog with hostile intentions"

"Dog in a humble and affectionate frame of mind"

"Half-bred shepherd dog"

"Dog caressing his master"

Anecdotal Approach

Evidence for emotions in animals has been primarily anecdotal, from individuals who interact with pets or captive animals on a regular basis. However, critics of animals having emotions often suggest that anthropomorphism is a motivating factor in the interpretation of the observed behaviours. Much of the debate is caused by the difficulty in defining emotions and the cognitive requirements thought necessary for animals to experience emotions in a similar way to humans. The problem is made more problematic by the difficulties in testing for emotions in animals. What is known about human emotion is almost all related or in relation to human communication.

Scientific Approach

In recent years, the scientific community has become increasingly supportive of the idea of emotions in animals. Scientific research has provided insight into similarities of physiological changes between humans and animals when experiencing emotion.

Much support for animal emotion and its expression results from the notion that feeling emotions doesn't require significant cognitive processes, rather, they could be motivated by the processes to act in an adaptive way, as suggested by Darwin. Recent attempts in studying emotions in animals have led to new constructions in experimental and information gathering. Professor Marian Dawkins suggested that emotions could be studied on a functional or a mechanistic basis. Dawkins suggests that merely mechanistic or functional research will provide the answer on its own, but suggests that a mixture of the two would yield the most significant results.

Functional

Functional approaches rely on understanding what roles emotions play in humans and examining that role in animals. A widely used framework for viewing emotions in a functional context is that described by Oatley and Jenkins who see emotions as having three stages: (i) appraisal in which there is a conscious or unconscious evaluation of an event as relevant to a particular goal. An emotion is positive when that goal is advanced and negative when it is impeded (ii) action readiness where the emotion gives priority to one or a few kinds of action and may give urgency to one so that it can interrupt or compete with others and (iii) physiological changes, facial expression and then behavioural action. The structure, however, may be too broad and could be used to include all the animal kingdom as well as some plants.

Mechanistic

The second approach, mechanistic, requires an examination of the mechanisms that drive emotions and search for similarities in animals.

The mechanistic approach is utilized extensively by Paul, Harding and Mendl. Recognizing the difficulty in studying emotion in non-verbal animals, Paul et al. demonstrate possible ways to better examine this. Observing the mechanisms that function in human emotion expression, Paul et al. suggest that concentration on similar mechanisms in animals can provide clear insights into the animal experience. They noted that in humans, cognitive biases vary according to emotional state and suggested this as a possible starting point to examine animal emotion. They propose that researchers may be able to use controlled stimuli which have a particular meaning to trained animals to induce particular emotions in these animals and assess which types of basic emotions animals can experience.

Cognitive Bias Test

Is the glass half empty or half full?

A cognitive bias is a pattern of deviation in judgment, whereby inferences about other animals and situations may be drawn in an illogical fashion. Individuals create their own "subjective social reality" from their perception of the input. It refers to the question "Is the glass half empty or half full?", used as an indicator of optimism or pessimism. To test this in animals, an individual is trained to anticipate that stimulus A, e.g. a 20 Hz tone, precedes a positive event, e.g. highly desired food is delivered when a lever is pressed by the animal. The same individual is trained to anticipate that stimulus B, e.g. a 10 Hz tone, precedes a negative event, e.g. bland food is delivered when the animal presses a lever. The animal is then tested by being played an intermediate stimulus C, e.g. a 15 Hz tone, and observing whether the animal presses the lever associated with the positive or negative reward, thereby indicating whether the animal is in a positive or negative mood. This might be influenced by, for example, the type of housing the animal is kept in.

Using this approach, it has been found that rats which are subjected to either handling or playful, experimenter-administered manual stimulation (tickling) showed different responses to the intermediate stimulus: rats exposed to tickling were more optimistic. The authors stated that they had demonstrated "for the first time a link between the directly measured positive affective state and decision making under uncertainty in an animal model."

Cognitive biases have been shown in a wide range of species including rats, dogs, rhesus macaques, sheep, chicks, starlings and honeybees.

Self-medication with Psychoactive Drugs

Humans can suffer from a range of emotional or mood disorders such as depression, anxiety, fear and panic. To treat these disorders, scientists have developed a range of psychoactive drugs such as anxiolytics. Many of these drugs are developed and tested by using a range of laboratory species. It is inconsistent to argue that these drugs are effective in treating human emotions whilst denying the experience of these emotions in the laboratory animals on which they have been developed and tested.

Standard laboratory cages prevent mice from performing several natural behaviours for which they are highly motivated. As a consequence, laboratory mice sometimes develop abnormal behaviours indicative of emotional disorders such as depression and anxiety. To improve welfare, these cages are sometimes enriched with items such as nesting material, shelters and running wheels. Sherwin and Ollson tested whether such enrichment influenced the consumption of Midazolam, a drug widely used to treat anxiety in humans. Mice in standard cages, standard cages but with unpredictable husbandry, or enriched cages, were given a choice of drinking either non-drugged water or a solution of the Midazolam. Mice in the standard and unpredictable cages drank a greater proportion of the anxiolytic solution than mice from enriched cages, indicating that mice from the standard and unpredictable laboratory caging may have been experiencing greater anxiety than mice from the enriched cages.

Spindle Neurons

Spindle neurons are specialised cells found in three very restricted regions of the human brain – the anterior cingulate cortex, the frontoinsular cortex and the dorsolateral prefrontal cortex. The first two of these areas regulate emotional functions such as empathy, speech, intuition, rapid "gut reactions" and social organization in humans. Spindle neurons are also found in the brains of humpback whales, fin whales, killer whales, sperm whales, bottlenose dolphin, Risso's dolphin, beluga whales, and the African and Asian elephants.

Whales have spindle cells in greater numbers and are maintained for twice as long as humans. The exact function of spindle cells in whale brains is still not understood, but Hof and Van Der Gucht believe that they act as some sort of "high-speed connections that fast-track information to and from other parts of the cortex". They compared them to express trains that bypass unnecessary connections, enabling organisms to instantly process and act on emotional cues during complex social interactions. However, Hof and Van Der Gucht clarify that they do not know the nature of such feelings in these animals and that we cannot just apply what we see in great apes or ourselves to whales. They believe that more work is needed to know whether emotions are the same for humans and whales.

Vocalisations

From at least the time of Darwin, it has been known that chimpanzees and other great apes also perform a laugh-like vocalization.

Research with rats has revealed that under particular conditions, they emit 50-kHz ultrasonic vocalisations (USV) which have been postulated to reflect a positive affective state (emotion) analogous to primitive human joy; these calls have been termed "laughter". The 50 kHz USVs in rats are uniquely elevated by hedonic stimuli—such as tickling, rewarding electrical brain stimulation, amphetamine injections, mating, play, and aggression—and are suppressed by aversive stimuli. Of all manipulations that elicit 50 kHz chirps in rats, tickling by humans elicits the highest rate of these calls.

Some vocalizations of domestic cats, such as purring, are well-known to be produced in situations of positive valence, such as mother kitten interactions, contacts with familiar partner, or during tactile stimulation with inanimate objects as when rolling and rubbing. Therefore, purring can be generally considered as an indicator of "pleasure" in cats.

Low pitched bleating in sheep has been associated with some positive-valence situations, as they are produced by males as an estrus female is approaching or by lactating mothers while licking and nursing their lambs.

Criticism

The argument that animals experience emotions is sometimes rejected due to a lack of evidence, and those who don't believe in the idea of animal intelligence, often argue that anthropomorphism plays a role in individuals' perspectives. Those who reject that animals have the capacity to experience emotion do so mainly by referring to inconsistencies in studies that have endorsed the belief emotions exist. Having no linguistic means to communicate emotion beyond behavioral response interpretation, the difficulty of providing an account of emotion in animals relies heavily on interpretive experimentation, that relies on results from human subjects.

Some people oppose the concept of animal emotions and suggest that emotions aren't universal, including in humans. If emotions are not universal, this indicates that there is not a phylogenetic relationship between human and non-human emotion. The relationship drawn by proponents of animal emotion, then, would be merely a suggestion of mechanistic features that promote adaptivity, but lack the complexity of human emotional constructs. Thus, a social life-style may play a role in the process of basic emotions developing into more complex emotions.

Darwin concluded, through a survey, that humans share universal emotive expressions and suggested that animals likely share in these to some degree.Social constructionists disregard the concept that emotions are universal. Others hold an intermediate stance, suggesting that basic emotional expressions and emotion are universal but the intricacies are developed culturally. A study by Elfenbein and Ambady indicated that individuals within a particular culture are better at recognising other cultural members' emotions.

Examples

Primates

Primates, in particular great apes, are candidates for being able to experience empathy and theory of mind. Great apes have complex social systems; young apes and their mothers have strong bonds of attachment and when a baby chimpanzee or gorilla dies, the mother will not uncommonly carry the body around for several days. Jane Goodall has described chimpanzees as exhibiting mournful behavior. Koko, a gorilla trained to use sign language, was reported to have expressed vocalisations indicating sadness after the death of her pet cat, All Ball.

Beyond such anecdotal evidence, support for empathetic reactions has come from experimental studies of rhesus macaques. Macaques refused to pull a chain that delivered food to themselves if doing so also caused a companion to receive an electric shock. This inhibition of hurting another conspecific was more pronounced between familiar than unfamiliar macaques, a finding similar to that of empathy in humans.

Furthermore, there has been research on consolation behavior in chimpanzees. De Waal and Aureli found that third-party contacts attempt to relieve the distress of contact participants by consoling (e.g. making contact, embracing, grooming) recipients of aggression, especially those that have experienced more intense aggression. Researchers were unable to replicate these results using the same observation protocol in studies of monkeys, demonstrating a possible difference in empathy between monkeys and apes.

Other studies have examined emotional processing in the great apes. Specifically, chimpanzees were shown video clips of emotionally charged scenes, such as a detested veterinary procedure or a favorite food, and then were required to match these scenes with one of two species-specific facial expressions: "happy" (a play-face) or "sad" (a teeth-baring expression seen in frustration or after defeat). The chimpanzees correctly matched the clips to the facial expressions that shared their meaning, demonstrating that they understand the emotional significance of their facial expressions. Measures of peripheral skin temperature also indicated that the video clips emotionally affected the chimpanzees.

Rodents

In 1998, Jaak Panksepp proposed that all mammalian species are equipped with brains capable of generating emotional experiences. Subsequent work examined studies on rodents to provide foundational support for this claim. One of these studies examined whether rats would work to alleviate the distress of a conspecific. Rats were trained to press a lever to avoid the delivery of an electric shock, signaled by a visual cue, to a conspecific. They were then tested in a situation in which either a conspecific or a Styrofoam block was hoisted into the air and could be lowered by pressing a lever. Rats that had previous experience with conspecific distress demonstrated greater than ten-fold more responses to lower a distressed conspecific compared to rats in the control group, while those who had never experienced conspecific distress expressed greater than three-fold more responses to lower a distressed conspecific relative to the control group. This suggests that rats will actively work to reduce the distress of a conspecific, a phenomenon related to empathy. Comparable results have also been found in similar experiments designed for monkeys.

Langford et al. examined empathy in rodents using an approach based in neuroscience. They reported that (1) if two mice experienced pain together, they expressed greater levels of pain-related behavior than if pain was experienced individually, (2) if experiencing different levels of pain together, the behavior of each mouse was modulated by the level of pain experienced by its social partner, and (3) sensitivity to a noxious stimulus was experienced to the same degree by the mouse observing a conspecific in pain as it was by the mouse directly experiencing the painful stimulus. The authors suggest this responsiveness to the pain of others demonstrated by mice is indicative of emotional contagion, a phenomenon associated with empathy, which has also been reported in pigs. One behaviour associated with fear in rats is freezing. If female rats experience electric shocks to the feet and then witness another rat experiencing similar footshocks, they freeze more than females without any experience of the shocks. This suggests empathy in experienced rats witnessing another individual being shocked. Furthermore, the demonstrator's behaviour was changed by the behaviour of the witness; demonstrators froze more following footshocks if their witness froze more creating an empathy loop.

Several studies have also shown rodents can respond to a conditioned stimulus that has been associated with the distress of a conspecific, as if it were paired with the direct experience of an unconditioned stimulus. These studies suggest that rodents are capable of shared affect, a concept critical to empathy.

Horses

Although not direct evidence that horses experience emotions, a 2016 study showed that domestic horses react differently to seeing photographs of positive (happy) or negative (angry) human facial expressions. When viewing angry faces, horses look more with their left eye which is associated with perceiving negative stimuli. Their heart rate also increases more quickly and they show more stress-related behaviours. One rider wrote, 'Experienced riders and trainers can learn to read the subtle moods of individual horses according to wisdom passed down from one horseman to the next, but also from years of trial-and-error. I suffered many bruised toes and nipped fingers before I could detect a curious swivel of the ears, irritated flick of the tail, or concerned crinkle above a long-lashed eye.' This suggests that horses have emotions and display them physically but it isn't concrete evidence.

Birds

Marc Bekoff reported accounts of animal behaviour which he believed was evidence of animals being able to experience emotions in his book *The Emotional Lives of Animals*. The following is an excerpt from his book:

> "A few years ago my friend Rod and I were riding our bicycles around Boulder, Colorado, when we witnessed a very interesting encounter among five magpies. Magpies are corvids, a very intelligent family of birds. One magpie had obviously been hit by a car and was laying dead on the side of the road. The four other magpies were standing around him. One approached the corpse, gently pecked at it-just as an elephant noses the carcass of another elephant- and stepped back. Another magpie did the same thing. Next, one of the magpies flew off, brought back some grass, and laid it by the corpse. Another magpie did the same. Then, all four magpies stood vigil for a few seconds and one by one flew off."

Bystander affiliation is believed to represent an expression of empathy in which the bystander tries to console a conflict victim and alleviate their distress. There is evidence for bystander affiliation in ravens (e.g. contact sitting, preening, or beak-to-beak or beak-to-body touching) and also for solicited bystander affiliation, in which there is post-conflict affiliation from the victim to the bystander. This indicates that ravens may be sensitive to the emotions of others, however, relationship value plays an important role in the prevalence and function of these post-conflict interactions.

The capacity of domestic hens to experience empathy has been studied. Mother hens show one of the essential underpinning attributes of empathy: the ability to be affected by, and share, the emotional state of their distressed chicks. However, evidence for empathy between familiar adult hens has not yet been found.

Dogs

A drawing by Konrad Lorenz showing facial expressions of a dog

Some research indicates that domestic dogs may experience negative emotions in a similar manner to humans, including the equivalent of certain chronic and acute psychological conditions. Much of this is from studies by Martin Seligman on the theory of learned helplessness as an extension of his interest in depression:

A dog that had earlier been repeatedly conditioned to associate an audible stimulus with inescapable electric shocks did not subsequently try to escape the electric shocks after the warning was presented, even though all the dog would have had to do is jump over a low divider within ten seconds. The dog didn't even try to avoid the "aversive stimulus"; it had previously "learned" that nothing it did would reduce the probability of it receiving a shock. A follow-up experiment involved three dogs affixed in harnesses, including one that received shocks of identical intensity and duration to the others, but the lever which would otherwise have allowed the dog a degree of control was left disconnected and didn't do anything. The first two dogs quickly recovered from the experience, but the third dog suffered chronic symptoms of clinical depression as a result of this perceived helplessness.

A further series of experiments showed that, similar to humans, under conditions of long-term intense psychological stress, around one third of dogs do not develop learned helplessness or long term depression. Instead these animals somehow managed to find a way to handle the unpleasant situation in spite of their past experience. The corresponding characteristic in humans has been found to correlate highly with an explanatory style and optimistic attitude that views the situation as *other than* personal, pervasive, or permanent.

Since these studies, symptoms analogous to clinical depression, neurosis, and other psychological conditions have also been accepted as being within the scope of emotion in domestic dogs. The postures of dogs may indicate their emotional state. In some instances, the recognition of specific postures and behaviors can be learned.

Psychology research has shown that when humans gaze at the face of another human, the gaze is not symmetrical; the gaze instinctively moves to the right side of the face to obtain information about their emotions and state. Research at the University of Lincoln shows that dogs share this instinct when meeting a human, and only when meeting a human (i.e. not other animals or other dogs). They are the only non-primate species known to share this instinct.

The existence and nature of personality traits in dogs have been studied (15,329 dogs of 164 different breeds). Five consistent and stable "narrow traits" were identified, described as playfulness, curiosity/fearlessness, chase-proneness, sociability and aggressiveness. A further higher order axis for shyness–boldness was also identified.

Dogs presented with images of either human or dog faces with different emotional states (happy/playful or angry/aggressive) paired with a single vocalization (voices or barks) from the same individual with either a positive or negative emotional state or brown noise. Dogs look longer at the face whose expression is congruent to the emotional state of the vocalization, for both other dogs and humans. This is an ability previously known only in humans. The behavior of a dog can not always be an indication its friendliness. This is because when a dog wags its tail, most people interpret this as the dog expressing happiness and friendliness. Though indeed tail wagging can express these positive emotions, tail wagging is also an indication of fear, insecurity, challenging of dominance, establishing social relationships or a warning that the dog may bite.

Some researchers are beginning to investigate the question of whether dogs have emotions with the help of magnetic resonance imaging.

Cats

It has been postulated that domestic cats can learn to manipulate their owners through vocalizations that are similar to the cries of human babies. Some cats learn to add a purr to the vocalization, which makes it less harmonious and more dissonant to humans, and therefore harder to ignore. Individual cats learn to make these vocalizations through trial-and-error; when a particular vocalization elicits a positive response from a human, the probability increases that the cat will use that vocalization in the future.

Growling can be an expression of annoyance or fear, similar to humans. When annoyed or angry, a cat wriggles and thumps its tail much more vigorously than when in a contented state. In larger felids such as lions, what appears to be irritating to them varies between individuals. A male lion may let his cubs play with his mane or tail, or he may hiss and hit them with his paws. Domestic male cats also have variable attitudes towards their family members, for example, older male siblings tend not to go near younger or new siblings and may even show hostility toward them.

Honeybees

Honeybees become pessimistic after being shaken

Honeybees ("Apis mellifera carnica") were trained to extend their proboscis to a two-component odour mixture (CS+) predicting a reward (e.g., 1.00 or 2.00 M sucrose) and to withhold their proboscis from another mixture (CS−) predicting either punishment or a less valuable reward (e.g., 0.01 M quinine solution or 0.3 M sucrose). Immediately after training, half of the honeybees were subjected to vigorous shaking for 60 s to simulate the state produced by a predatory attack on a concealed colony. This shaking reduced levels of octopamine, dopamine, and serotonin in the hemolymph of a separate group of honeybees at a time point corresponding to when the cognitive bias tests were performed. In honeybees, octopamine is the local neurotransmitter that functions during reward learning, whereas dopamine mediates the ability to learn to associate odours with quinine punishment. If flies are fed serotonin, they are more aggressive; flies depleted of serotonin still exhibit aggression, but they do so much less frequently.

Within 5 minutes of the shaking, all the trained bees began a sequence of unreinforced test trials with five odour stimuli presented in a random order for each bee: the CS+, the CS−, and three novel odours composed of ratios intermediate between the two learned mixtures. Shaken honeybees were more likely to withhold their mouthparts from the CS− and from the most similar novel odour. Therefore, agitated honeybees display an increased expectation of bad outcomes similar to a vertebrate-like emotional state. The researchers of the study stated that, "Although our results do not allow us to make any claims about the presence of negative subjective feelings in honeybees, they call into question how we identify emotions in any non-human animal. It is logically inconsistent to claim that the presence of pessimistic cognitive biases should be taken as confirmation that dogs or rats are anxious but to deny the same conclusion in the case of honeybees."

Crayfish

Crayfish naturally explore new environments but display a general preference for dark places. A 2014 study on the freshwater crayfish *Procambarus clarkii* tested their responses in a fear paradigm, the elevated plus maze in which animals choose to walk on an elevated cross which offers both aversive and preferable conditions (in this case, two arms were lit and two were dark). Crayfish which experienced an electric shock displayed enhanced fearfulness or anxiety as demonstrated by their preference for the dark arms more than the light. Furthermore, shocked crayfish had relatively higher brain serotonin concentrations coupled with elevated blood glucose, which indicates a stress response. Moreover, the crayfish calmed down when they were injected with the benzodiazepine anxiolytic, chlordiazepoxide, used to treat anxiety in humans, and they entered the dark as normal. The authors of the study concluded "stress-induced avoidance behavior in crayfish exhibits striking homologies with vertebrate anxiety."

The freshwater crayfish *Procambarus clarkii*

A follow-up study using the same species confirmed the anxiolytic effect of chlordiazepoxide, but moreover, the intensity of the anxiety-like behaviour was dependent on the intensity of the electric shock until reaching a plateau. Such a quantitative relationship between stress and anxiety is also a very common feature of human and vertebrate anxiety.

Pain in Animals

In humans, pain is a distressing feeling often caused by intense or damaging stimuli. Whether animals apart from humans also experience pain is often contentious despite being scientifically verifiable. The standard measure of pain in humans is how a person reports that pain, (for example, on a pain scale). "Pain" is defined by the International Association for the Study of Pain as "an unpleasant sensory and emotional experience associated with actual or potential tissue damage, or described in terms of such damage." Only the person experiencing the pain can know the pain's quality and intensity, and the degree of suffering. However, for non-human animals, it is harder, if even possible, to know whether an emotional experience has occurred. Therefore, this concept is often excluded in definitions of pain in animals, such as that provided by Zimmerman: "an aversive sensory experience caused by actual or potential injury that elicits protective motor and vegetative reactions, results in learned avoidance and may modify species-specific behaviour, including social behaviour." Non-human animals cannot report their feelings to language-using humans in the same manner as human communication, but observation of their behaviour provides a reasonable indication as to the extent of their pain. Just as with doctors and medics who sometimes share no common language with their patients, the indicators of pain can still be understood.

A Galapagos shark hooked by a fishing boat

According to the U.S. National Research Council Committee on Recognition and Alleviation of Pain in Laboratory Animals, pain is experienced by many animal species, including mammals and possibly all vertebrates.

The Experience of Pain

Although there are numerous definitions of pain, almost all involve two key components. First, nociception is required. This is the ability to detect noxious stimuli which evoke a reflex response that rapidly moves the entire animal, or the affected part of its body, away from the source of the stimulus. The concept of nociception does not imply any adverse, subjective "feeling" – it is a reflex action. An example in humans would be the rapid withdrawal of a finger that has touched something hot – the withdrawal occurs before any sensation of pain is actually experienced.

The second component is the experience of "pain" itself, or suffering – the internal, emotional interpretation of the nociceptive experience. Again in humans, this is when the withdrawn finger begins to hurt, moments after the withdrawal. Pain is therefore a private, emotional experience. Pain cannot be directly measured in other animals, including other humans; responses to putatively painful stimuli can be measured, but not the experience itself. To address this problem when assessing the capacity of other species to experience pain, argument-by-analogy is used. This is based on the principle that if an animal responds to a stimulus in a similar way to ourselves, it is likely to have had an analogous experience.

Reflex Response to Painful Stimuli

Nociception usually involves the transmission of a signal along nerve fibers from the site of a noxious stimulus at the periphery to the spinal cord. Although this signal is also transmitted on to the brain, a reflex response, such as flinching or withdrawal of a limb, is produced by return signals originating in the spinal cord. Thus, both physiological and behavioral responses to nociception can be detected, and no reference need be made to a conscious experience of pain. Based such criteria nociception has been observed in all major animal taxa.

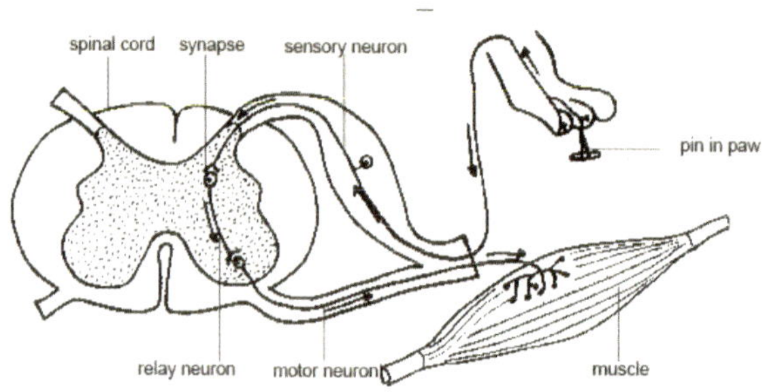

Reflex arc of a dog when its paw is stuck with a pin. The spinal cord responds to signals from receptors in the paw, producing a reflex withdrawal of the paw. This localized response does not involve brain processes that might mediate a consciousness of pain, though these might also occur

Awareness of Pain

Nerve impulses from nociceptors may reach the brain, where information about the stimulus (e.g. quality, location, and intensity), and affect (unpleasantness) are registered. Though the brain activity involved has been studied, the brain processes underlying conscious awareness are not well understood.

Adaptive Value

The adaptive value of nociception is obvious; an organism detecting a noxious stimulus immediately withdraws the limb, appendage or entire body from the noxious stimulus and thereby avoids further (potential) injury. However, a characteristic of pain (in mammals at least) is that pain can result in hyperalgesia (a heightened sensitivity to noxious stimuli) and allodynia (a heightened sensitivity to non-noxious stimuli). When this heightened sensitisation occurs, the adaptive value is less clear. First, the pain arising from the heightened sensitisation can be disproportionate to the actual tissue damage caused. Second, the heightened sensitisation may also become chronic, persisting well beyond the tissues healing. This can mean that rather than the actual tissue damage causing pain, it is the pain due to the heightened sensitisation that becomes the concern. This means the sensitisation process is sometimes termed maladaptive. It is often suggested hyperalgesia and allodynia assist organisms to protect themselves during healing, but experimental evidence to support this has been lacking.

In 2014, the adaptive value of sensitisation due to injury was tested using the predatory interactions between longfin inshore squid (*Doryteuthis pealeii*) and black sea bass (*Centropristis striata*) which are natural predators of this squid. If injured squid are targeted by a bass, they began their defensive behaviours sooner (indicated by greater alert distances and longer flight initiation distances) than uninjured squid. If anaesthetic (1% ethanol and $MgCl_2$) is administered prior to the injury, this prevents the sensitisation and blocks the behavioural effect. The authors claim this study is the first experimental evidence to support the argument that nociceptive sensitisation is actually an adaptive response to injuries.

Argument-by-analogy

To assess the capacity of other species to consciously suffer pain we resort to argument-by-analogy. That is, if an animal responds to a stimulus the way a human does, it is likely to have had an analogous experience. If we stick a pin in a chimpanzee's finger and she rapidly withdraws her hand, we use argument-by-analogy and infer that like us, she felt pain. It might be argued that consistency requires us infer, also, that a cockroach experiences conscious pain when it writhes after being stuck with a pin. The usual counter-argument is that although the physiology of consciousness is not understood, it clearly involves complex brain processes not present in relatively simple organisms. Other analogies have been pointed out. For example, when given a choice of foods, rats and chickens with clinical symptoms of pain will consume more of an analgesic-containing food than animals not in pain. Additionally, the consumption of the analgesic carprofen in lame chickens was positively correlated to the severity of lameness, and consumption resulted in an improved gait. Such anthropomorphic arguments face the criticism that physical reactions indicating pain may be neither the cause nor result of conscious states, and the approach is subject to criticism of anthropomorphic interpretation. For example, a single-celled organism such as an amoeba may writhe after being exposed to noxious stimuli despite the absence of nociception.

In Different Species

The ability to experience pain in an animal, or another human for that matter, cannot be determined directly but it may be inferred through analogous physiological and behavioral reactions. Although many animals share similar mechanisms of pain detection to those of humans, have similar areas of the brain involved in processing pain, and show similar pain behaviours, it is notoriously difficult to assess how animals actually experience pain.

Nociception

Nociceptive nerves, which preferentially detect (potential) injury-causing stimuli, have been identified in a variety of animals, including invertebrates. The medicinal leech, *Hirudo medicinalis,* and sea slug are classic model systems for studying nociception. Many other vertebrate and invertebrate animals also show nociceptive reflex responses similar to our own.

Pain

Many animals also exhibit more complex behavioural and physiological changes indicative of the ability to experience pain: they eat less food, their normal behaviour is disrupted, their social behaviour is suppressed, they may adopt unusual behaviour patterns, they may emit characteristic distress calls, experience respiratory and cardiovascular changes, as well as inflammation and release of stress hormones.

Some criteria that may indicate the potential of another species to feel pain include:

1. Has a suitable nervous system and sensory receptors
2. Physiological changes to noxious stimuli
3. Displays protective motor reactions that might include reduced use of an affected area such as limping, rubbing, holding or autotomy
4. Has opioid receptors and shows reduced responses to noxious stimuli when given analgesics and local anaesthetics

5. Shows trade-offs between stimulus avoidance and other motivational requirements

6. Shows avoidance learning

7. High cognitive ability and sentience

Vertebrates

Fish

A typical human cutaneous nerve contains 83% C type trauma receptors (the type responsible for transmitting signals described by humans as excruciating pain); the same nerves in humans with congenital insensitivity to pain have only 24-28% C type receptors. The rainbow trout has about 5% C type fibres, while sharks and rays have 0%. Nevertheless, fish have been shown to have sensory neurons that are sensitive to damaging stimuli and are physiologically identical to human nociceptors. Behavioural and physiological responses to a painful event appear comparable to those seen in amphibians, birds, and mammals, and administration of an analgesic drug reduces these responses in fish.

Animal welfare advocates have raised concerns about the possible suffering of fish caused by angling. Some countries, e.g. Germany, have banned specific types of fishing, and the British RSPCA now formally prosecutes individuals who are cruel to fish.

Invertebrates

Though it has been argued that most invertebrates do not feel pain, there is some evidence that invertebrates, especially the decapod crustaceans (e.g. crabs and lobsters) and cephalopods (e.g. octopuses), exhibit behavioural and physiological reactions indicating they may have the capacity for this experience. Nociceptors have been found in nematodes, annelids and molluscs. Most insects do not possess nociceptors, one known exception being the fruit fly. In vertebrates, endogenous opioids are neurochemicals that moderate pain by interacting with opiate receptors. Opioid peptides and opiate receptors occur naturally in nematodes, molluscs, insects and crustaceans. The presence of opioids in crustaceans has been interpreted as an indication that lobsters may be able to experience pain, although it has been claimed "at present no certain conclusion can be drawn".

One suggested reason for rejecting a pain experience in invertebrates is that invertebrate brains are too small. However, brain size does not necessarily equate to complexity of function. Moreover, weight for body-weight, the cephalopod brain is in the same size bracket as the vertebrate brain, smaller than that of birds and mammals, but as big as or bigger than most fish brains.

Since September 2010, all cephalopods being used for scientific purposes in the EU are protected by EU Directive 2010/63/EU which states "there is scientific evidence of their [cephalopods] ability to experience pain, suffering, distress and lasting harm. In the UK, animal protection legislation means that cephalopods used for scientific purposes must be killed humanely, according to prescribed methods (known as "Schedule 1 methods of euthanasia") known to minimise suffering.

In Medicine and Research

Veterinary Medicine

Veterinary medicine uses, for actual or potential animal pain, the same analgesics and anesthetics as used in humans.

Dolorimetry

Dolorimetry (*dolor*: Latin: pain, grief) is the measurement of the pain response in animals, including humans. It is practiced occasionally in medicine, as a diagnostic tool, and is regularly used in research into the basic science of pain, and in testing the efficacy of analgesics. Non-human animal pain measurement techniques include the paw pressure test, tail flick test, hot plate test and grimace scales.

Laboratory Animals

Animals are kept in laboratories for a wide range of reasons, some of which may involve pain, suffering or distress, whilst others (e.g. many of those involved in breeding) will not. The extent to which animal testing causes pain and suffering in laboratory animals is the subject of much debate. Marian Stamp Dawkins defines "suffering" in laboratory animals as the experience of one of "a wide range of extremely unpleasant subjective (mental) states." The U.S. National Research Council has published guidelines on the care and use of laboratory animals, as well as a report on recognizing and alleviating pain in vertebrates. The United States Department of Agriculture defines a "painful procedure" in an animal study as one that would "reasonably be expected to cause more than slight or momentary pain or distress in a human being to which that procedure was applied." Some critics argue that, paradoxically, researchers raised in the era of increased awareness of animal welfare may be inclined to deny that animals are in pain simply because they do not want to see themselves as people who inflict it. PETA however argues that there is no doubt about animals in laboratories being inflicted with pain. In the UK, animal research likely to cause "pain, suffering, distress or lasting harm" is regulated by the Animals (Scientific Procedures) Act 1986 and research with the potential to cause pain is regulated by the Animal Welfare Act of 1966 in the US.

In the U.S., researchers are not required to provide laboratory animals with pain relief if the administration of such drugs would interfere with their experiment. Laboratory animal veterinarian Larry Carbone writes, "Without question, present public policy allows humans to cause laboratory animals unalleviated pain. The AWA, the *Guide for the Care and Use of Laboratory Animals*, and current Public Health Service policy all allow for the conduct of what are often called "Category E" studies – experiments in which animals are expected to undergo significant pain or distress that will be left untreated because treatments for pain would be expected to interfere with the experiment."

Severity Scales

Eleven countries have national classification systems of pain and suffering experienced by animals used in research: Australia, Canada, Finland, Germany, The Republic of Ireland, The Netherlands,

New Zealand, Poland, Sweden, Switzerland, and the UK. The US also has a mandated national scientific animal-use classification system, but it is markedly different from other countries in that it reports on whether pain-relieving drugs were required and/or used. The first severity scales were implemented in 1986 by Finland and the UK. The number of severity categories ranges between 3 (Sweden and Finland) and 9 (Australia). In the UK, research projects are classified as "mild", "moderate", and "substantial" in terms of the suffering the researchers conducting the study say they may cause; a fourth category of "unclassified" means the animal was anesthetized and killed without recovering consciousness. It should be remembered that in the UK system, many research projects (e.g. transgenic breeding, feeding distasteful food) will require a license under the Animals (Scientific Procedures) Act 1986, but may cause little or no pain or suffering. In December 2001, 39 percent (1, 296) of project licenses in force were classified as "mild", 55 percent (1, 811) as "moderate", two percent (63) as "substantial", and 4 percent (139) as "unclassified". In 2009, of the project licenses issued, 35 percent (187) were classified as "mild", 61 percent (330) as "moderate", 2 percent (13) as "severe" and 2 percent (11) as unclassified.

In the US, the *Guide for the Care and Use of Laboratory Animals* defines the parameters for animal testing regulations. It states, "The ability to experience and respond to pain is widespread in the animal kingdom...Pain is a stressor and, if not relieved, can lead to unacceptable levels of stress and distress in animals. " The *Guide* states that the ability to recognize the symptoms of pain in different species is essential for the people caring for and using animals. Accordingly, all issues of animal pain and distress, and their potential treatment with analgesia and anesthesia, are required regulatory issues for animal protocol approval.

Tool use by Animals

A crab-eating macaque using a stone

Tool use by animals is a phenomenon in which an animal uses any kind of tool in order to achieve a goal such as acquiring food and water, grooming, defense, recreation or construction. Originally thought to be a skill only possessed by humans, some tool use requires a sophisticated level of cognition. There is considerable discussion about the definition of what constitutes a tool and therefore which behaviours can be considered as true examples of tool use. A wide range of animals are considered to use tools including mammals, birds, fish, cephalopods and insects.

Primates are well known for using tools for hunting or gathering food and water, cover for rain, and self-defence. Chimpanzees have been the object of study, most famously by Jane Goodall, since these animals are more-often kept in captivity than other primates and are closely related to humans. Tool-use in other primates are lesser-known as many of them are mainly observed in the wild. Many famous researchers, such as Charles Darwin in his book *The Descent of Man*, mentioned tool-use in monkeys (such as baboons). Both wild and captive elephants are known to create tools using their trunk and feet, mainly for swatting flies, scratching, plugging water-holes (so the water doesn't evaporate), and reaching food that is out of reach. A group of dolphins in Shark Bay use sponges to protect their beak while foraging. Sea otters will dislodge food from rocks (such as abalone) and break open shellfish. Carnivores (of the order Carnivora) can use tools to trap prey or break open the shells of prey, as well as for scratching.

Corvids (crows, ravens and rooks) are well known for their large brains (among birds) and subsequent tool use. They mainly manufacture probes out of twigs and wood (and sometimes metal wire) to catch or impale larvae. Crows are among the only animals that create their own toys. Tool use in other birds is best exemplified in nest intricacy. Warblers manufacture 'pouches' to make their nests in. Some birds, such as weaver birds build complex nests. Finches and woodpeckers may insert twigs into trees in order to catch or impale larvae. Parrots may use tools to wedge nuts so that they may crack it open (using a tool) without launching it away. Some birds take advantage of human activity, such as some species of gulls which drop shellfish in front of cars to crack them open.

Several species of fish use tools to crack open shellfish, extract food that is out of reach, cleaning an area (for nesting), and hunting. Octopuses gather coconut shells and create a shelter. They may also construct a fence using rocks.

Definitions and Terminology

The key to identifying tool use is defining what constitutes a tool. Researchers of animal behavior have arrived at different formulations.

In 1980, Beck published a widely used definition of tool use. This has been modified to:

The external employment of an unattached or manipulable attached environmental object to alter more efficiently the form, position, or condition of another object, another organism, or the user itself, when the user holds and directly manipulates the tool during or prior to use and is responsible for the proper and effective orientation of the tool.

Other, briefer definitions have been proposed:

An object carried or maintained for future use.

—Finn, Tregenza, and Norman, 2009.

The use of physical objects other than the animal's own body or appendages as a means to extend the physical influence realized by the animal.

—Jones and Kamil, 1973

An object that has been modified to fit a purpose ... [or] An inanimate object that one uses or modifies in some way to cause a change in the environment, thereby facilitating one's achievement of a target goal.

— Hauser, 2000

Others, for example Lawick-Goodall, distinguish between "tool use" and "object use".

Different terms have been given to the tool according to whether the tool is altered by the animal. If the "tool" is not held or manipulated by the animal in any way, such as an immobile anvil, objects in a bowerbird's bower, or a bird using bread as bait to catch fish, it is sometimes referred to as a "proto-tool". Several studies in primates and birds have found that tool use is correlated with an enlargement of the brain as a whole or of particular regions. For example, true tool-using birds have relatively larger brains than proto-tool users.

When an animal uses a tool that acts on another tool, this has been termed use of a "meta-tool". For example, New Caledonian crows will spontaneously use a short tool to obtain an otherwise inaccessible longer tool that then allows them to extract food from a hole. Similarly, bearded capuchin monkeys will use smaller stones to loosen bigger quartz pebbles embedded in conglomerate rock, which they subsequently use as tools.

Rarely, animals may use one tool followed by another, for example, bearded capuchins use stones and sticks, or two stones. This is called "associative", "secondary" or "sequential" tool use.

Some animals use other individuals in a way which could be interpreted as tool use, for example, ants crossing water over a bridge of other ants, or weaver ants using conspecifics to glue leaves together. These have been termed "social tools".

Borderline Examples

Play: Play has been defined as "activity having no immediate benefits and structurally including repetitive or exaggerated actions that may be out of sequence or disordered". When play is discussed in relation to manipulating objects, it is often used in association with the word "tool". Some birds, notably crows, parrots and birds of prey, "play" with objects, many of them playing in flight with such items as stones, sticks and leaves, by letting them go and catching them again before they reach the ground. A few species repeatedly drop stones, apparently for the enjoyment of the sound effects. Many other species of animals, both avian and non-avian, play with objects in a similar manner.

Fixed "devices": The impaling of prey on thorns by many of the shrikes (*Laniidae*) is well known. Several other birds may use spines or forked sticks to anchor a carcass while they flay it with the bill. It has been concluded that "This is an example of a fixed device which serves as an extension of the body, in this case, talons" and is thus a true form of tool use. On the other hand, the use of fixed skewers may not be true tool-use because the thorn (or whatever) is not manipulated by the bird. Leopards perform a similar behaviour by dragging carcasses up trees and caching them in the forks of branches.

Use of bait: Several species of bird, including herons such as the striated heron (*Butorides striatus*), will place bread in water to attract fish. Whether this is tool use is disputed because the bread is not manipulated or held by the bird.

Learning and Cognition

Tool use by animals may indicate different levels of learning and cognition. For some animals, tool use is largely instinctive and inflexible. For example, the woodpecker finch of the Galápagos Islands use twigs or spines as an essential and regular part of its foraging behaviour, but these behaviours are often quite inflexible and are not applied effectively in different situations. Other tool use, e.g. chimpanzees using twigs to "fish" for termites, may be developed by watching others use tools and may even be a true example of animal teaching. Tools may even be used in solving puzzles in which the animal appears to experience a "Eureka moment".

Primates

A gorilla using a stick possibly to gauge the depth of water

Tool use has been reported many times in both wild and captive primates, particularly the great apes. The use of tools by primates is varied and includes hunting (mammals, invertebrates, fish), collecting honey, processing food (nuts, fruits, vegetables and seeds), collecting water, weapons and shelter.

In 1960, Jane Goodall observed a chimpanzee poking pieces of grass into a termite mound and then raising the grass to his mouth. After he left, Goodall approached the mound and repeated the behaviour because she was unsure what the chimpanzee was doing. She found that the termites bit onto the grass with their jaws. The chimpanzee had been using the grass as a tool to "fish" or "dip" for termites.

Tool manufacture is much rarer than simple tool use and probably represents higher cognitive functioning. Soon after her initial discovery of tool use, Goodall observed other chimpanzees picking up leafy twigs, stripping off the leaves and using the stems to fish for insects. This change of a leafy twig into a tool was a major discovery. Prior to this, scientists thought that only humans manufactured and used tools, and that this ability was what separated humans from other animals. In 1990, it was claimed the only primate to manufacture tools in the wild was the chimpanzee. However, since then, several primates have been reported as tool makers in the wild.

Both bonobos and chimpanzees have been observed making "sponges" out of leaves and moss that suck up water and using these for grooming. Sumatran orangutans will take a live branch, remove twigs and leaves and sometimes the bark, before fraying or flattening the tip for use on ants or bees. In the wild, mandrills have been observed to clean their ears with modified tools. Scientists filmed a large male mandrill at Chester Zoo (UK) stripping down a twig, apparently to make it narrower, and then using the modified stick to scrape dirt from underneath his toenails. Captive gorillas have made a variety of tools.

Chimpanzees and Bonobos

Common chimpanzees (*Pan troglodytes*) are sophisticated tool users with behaviors including cracking nuts with stone tools and fishing for ants or termites with sticks. There are more limited reports of the closely related bonobo (*Pan paniscus*) using tools in the wild; it has been claimed they rarely use tools in the wild although they use tools as readily as chimpanzees when in captivity, It has been reported that females of both chimpanzees and bonobos use tools more avidly than males. Wild chimpanzees predominantly use tools in the context of food acquisition, while wild bonobos appear to use tools mainly for personal care (cleaning, protection from rain) and social purposes. Wild bonobos have been observed using leaves as cover for rain, or the use of branches in social displays.

Hunting

Research in 2007 showed that common chimpanzees sharpen sticks to use as weapons when hunting mammals. This is considered the first evidence of systematic use of weapons in a species other than humans. Researchers documented 22 occasions when wild chimpanzees on a savanna in Senegal fashioned sticks into "spears" to hunt lesser bushbabies (*Galago senegalensis*). In each case, a chimpanzee modified a branch by breaking off one or two ends and, frequently using its teeth, sharpened the stick. The tools, on average, were about 60 cm (24 in) long and 1.1 cm (0.4 in) in circumference. The chimpanzee then jabbed the spear into hollows in tree trunks where bushbabies sleep. There was a single case in which a chimpanzee successfully extracted a bushbaby with the tool. It has been suggested that the word "spear" is an overstatement that makes the chimpanzees seem too much like early humans, and that the term "bludgeon" is more accurate, since the point of the tool may not be particularly sharp. This behaviour was seen more frequently in females, particularly adolescent females, and young chimps in general, than in adult males.

Chimpanzees often eat the marrow of long bones of colobus monkeys with the help of small sticks, after opening the ends of the bones with their teeth. A juvenile female was observed to eat small parts of the brain of an intact skull that she could not break open by inserting a small stick through the foramen magnum. On another occasion, an adult female used three sticks to clean the orbits of a colobus monkey skull after she had just eaten the eyes.

A Bonobo Fishing for Termites

In Gombe National Park in 1960, Jane Goodall observed a chimpanzee, David Greybeard, poking pieces of grass into a termite mound and then raising the grass to his mouth. After he left, Goodall approached the mound and repeated the behaviour because she was unsure what David was doing. She found that the termites bit onto the grass with their jaws. David had been using the grass as a tool to "fish" or "dip" for termites. Soon after this initial discovery of tool use, Goodall observed

David and other chimpanzees picking up leafy twigs, stripping off the leaves, and using the stems to fish for insects. This modification of a leafy twig into a tool was a major discovery: previously, scientists thought that only humans made and used tools, and that this was what separated humans from other animals.

Other studies of the Gombe chimps show that young females and males learn to fish for termites differently. Female chimps learn to fish for termites earlier and better than the young males. Females also spend more time fishing while at the mounds with their mothers—males spend more time playing. When they are adults, females need more termite protein because with young to care for, they cannot hunt the way males can.

Populations differ in the prevalence of tool use for fishing for invertebrates. Chimpanzees in the Tai National Park only sometimes use tools, whereas Gombe chimpanzees rely almost exclusively on tools for their intake of driver ants. This may be due to difference in the rewards gained by tool use: Gombe chimpanzees collect 760 ants/min compared to 180 ants/min for the Tai chimpanzees.

Some chimpanzees use tools to hunt large bees (*Xylocopa* sp.) which make nests in dead branches on the ground or in trees. To get to the grubs and the honey, the chimpanzee first tests for the presence of adults by probing the nest entrance with a stick. If present, adult bees block the entrance with their abdomens, ready to sting. The chimpanzee then disables them with the stick to make them fall out and eats them rapidly. Afterwards, the chimpanzee opens the branch with its teeth to obtain the grubs and the honey.

Chimpanzees have even been observed using two tools: a stick to dig into an ant nest and a "brush" made from grass stems with their teeth to collect the ants.

Collecting Honey

Honey of four bee species is eaten by chimpanzees. Groups of chimpanzees fish with sticks for the honey after having tried to remove what they can with their hands. They usually extract with their hands honeycombs from undisturbed hives of honey bees and run away from the bees to quietly eat their catch. In contrast, hives that have already been disturbed, either through the falling of the tree or because of the intervention of other predators, are cleaned of the remaining honey with fishing tools.

Processing Food

A chimpanzee gathering food with a stick

Tai chimpanzees crack open nuts with rocks, but there is no record of Gombe chimpanzees using rocks in this way. After opening nuts by pounding with a hammer, parts of the kernels may be too difficult to reach with the teeth or fingernails, and some individuals use sticks to remove these remains, instead of pounding the nut further with the hammer as other individuals do: a relatively rare combination of using two different tools. Hammers for opening nuts may be either wood or stone.

Chimpanzees in the Nimba Mountains of Guinea, Africa, use both stone and wooden cleavers, as well as stone anvils, to chop up and reduce *Treculia* fruits into smaller bite-sized portions. These fruits, which can be the size of a volleyball and weigh up to 8.5 kg, are hard and fibrous. But, despite lacking a hard outer shell, they are too large for a chimpanzee to get its jaws around and bite into. Instead, the chimpanzees use a range of tools to chop them into smaller pieces. This is the first account of chimpanzees using a pounding tool technology to break down large food items into bite-sized chunks rather than just extracting it from other unobtainable sources such as baobab nuts. It is also the first time wild chimpanzees have been found to use two distinct types of percussive technology, i.e. movable cleavers against a non-movable anvil, to achieve the same goal. Neighbouring chimpanzees in the nearby region of Seringbara do not process their food in this way, indicating how tool use among apes is culturally learnt.

Collecting Water

When chimpanzees cannot reach water that has formed in hollows high up inside trees, they have been observed taking a handful of leaves, chewing them, and dipping this "sponge" into the pool to suck out the water. Both bonobos and chimpanzees have also been observed making "sponges" out of leaves and moss that suck up water and are used as grooming tools.

Orangutans

Orangutans (genus *Pongo*) were first observed using tools in the wild in 1994 in the northwest corner of Sumatra. As with the chimpanzees, orangutans use tools made from branches and leaves to scratch, scrape, wipe, sponge, swat, fan, hook, probe, scoop, pry, chisel, hammer, cover, cushion and amplify. They will break off a tree branch that is about 30 cm long, snap off the twigs, fray one end and then use the stick to dig in tree holes for termites. Sumatran orangutans use a variety of tools—up to 54 types for extracting insects or honey, and as many as 20 types for opening or preparing fruits such as the hard to access *Neesia Malayana*. They also use an 'autoerotic tool'—a stick which they use to stimulate the genitals and masturbate (both male and female). In parts of Borneo, orangutans use handfuls of leaves as napkins to wipe their chins while orangutans in parts of Sumatra use leaves as gloves, helping them handle spiny fruits and branches, or as seat cushions in spiny trees. There have been reports that individuals in both captivity and in the wild use tools held between the lips or teeth, rather than in the hands. In captivity, orangutans have been taught to chip stone handaxes.

Orangutans living in Borneo scavenge fish that wash up along the shore and scoop catfish out of small ponds for fresh meals. Over two years, anthropologist Anne Russon saw several animals on these forested islands learn on their own to jab at catfish with sticks, so that the panicked prey would flop out of ponds and into the orangutan's waiting hands. Although orangutans usually fished alone, Russon observed pairs of apes catching catfish on a few occasions. On the island of Kaja in Borneo, a male orangutan was observed using a pole apparently trying to spear or bludgeon fish. This individual had seen humans fishing with spears. Although not successful, he was later able to improvise by using the pole to catch fish already trapped in the locals' fishing lines.

Sumatran orangutans use sticks to acquire seeds from a particular fruit. When the fruit of the *Neesia* tree ripens, its hard, ridged husk softens until it falls open. Inside are seeds that are highly desirable to the orangutans, but they are surrounded by fibreglass-like hairs that are painful if eaten. A *Neesia*-eating orangutan will select a 12 cm stick, strip off the bark, and then carefully collect the hairs with it. Once the fruit is safe, the ape will eat the seeds using the stick or its fingers. Sumatran orangutans will use a stick to poke a bees' nest wall, move it around and catch the honey.

Orangutans have been observed using sticks to apparently measure the depth of water. It has been reported that orangutans use tools for a wide range of purposes including using leaves as protective gloves or napkins, using leafy branches to swat insects or gather water, and building sun or rain covers above the nests used for resting. It has been reported that a Sumatran orangutan used a large leaf as an umbrella in a tropical rainstorm.

Orangutans produce an alarm call known as a "kiss squeak" when they encounter a predator like a snake or a human. Sometimes, orangutans will strip leaves from a branch and hold them in front of their mouth when making the sound. It has been found this lowers the maximum frequency of the sound i.e. makes it deeper, and in addition, smaller orangutans are more likely to use the leaves. It has been suggested they use the leaves to make themselves sound bigger than they really are, the first documented case of an animal using a tool to manipulate sound.

Gorillas

A gorilla pushing a stick into the ground and using it as a stabilizing stick while dredging aquatic herbs

There are few reports of gorillas using tools in the wild. Western lowland gorillas (*Gorilla gorilla gorilla*) have been observed using sticks to apparently measure the depth of water and as "walking sticks" to support their posture when crossing deeper water. An adult female used a detached trunk from a small shrub as a stabilizer during food gathering, and another used a log as a bridge. One possible explanation for the absence of observed tool use in wild gorillas is that they are less dependent on foraging techniques that require the use of tools, since they exploit food resources differently from chimpanzees. Whereas chimpanzees and orangutans feeding involves tools such as hammers to crack open nuts and sticks to fish for termites, gorillas access these foods by breaking nuts with their teeth and smashing termite mounds with their hands.

Captive western lowland gorillas have been observed to threaten each other with sticks and larger pieces of wood, while others use sticks for hygienic purposes. Some females have attempted to use logs as ladders. In another group of captive gorillas, several individuals were observed throwing sticks and branches into a tree, apparently to knock down leaves and seeds. Gorillas at Prague zoo have used tools in several ways, including using wood wool as "slippers" when walking on the snow or to cross a wet section of the floor.

Monkeys

Tool use has been observed in capuchin monkeys both in captivity and in their natural environments. In a captive environment, capuchins readily insert a stick into a tube containing viscous food that clings to the stick, which they then extract and lick. Capuchins also use a stick to push food from the center of a tube retrieving the food when it reaches the far end, and as a rake to sweep objects or food toward themselves. The black-striped capuchin (*Sapajus libidinosus*) was the first non-ape primate for which tool use was documented in the wild; individuals were observed cracking nuts by placing them on a stone anvil and hitting them with another large stone (hammer). Similar hammer-and-anvil use has been observed in other wild capuchins including robust capuchin monkeys (genus *Sapajus*) It may take a capuchin up to 8 years to master this skill. The monkeys often transport hard fruits, stones, nuts and even oysters to an anvil for this purpose. Capuchins also use stones as digging tools for probing the substrate and sometimes for excavating tubers. Wild black-striped capuchin use sticks to flush prey from inside rock crevices. Robust capuchins are also known at times to rub defensive secretions from arthropods over their bodies before eating them; such secretions are believed to act as natural insecticides.

Darwin mentioned tool use by wild baboons in *The Descent of Man*:

Brehm states, on the authority of the well-known traveller Schimper, that in Abyssinia when the baboons belonging to one species (*C. gelada*) descend in troops from the mountains to plunder the fields, they sometimes encounter troops of another species (*C. hamadryas*), and then a fight ensues. The Geladas roll down great stones, which the Hamadryas try to avoid...

Darwin continued:

Brehm, when accompanying the Duke of Coburg-Gotha, aided in an attack with fire-arms on a troop of baboons in the pass of Mensa in Abyssinia. The baboons in return rolled so many stones down the mountain, some as large as a man's head, that the attackers had to beat a hasty retreat; and the pass was actually for a time closed against the caravan.

These rather anecdotal reports of stone throwing by baboons have been corroborated by more recent research on chacma baboon (*Papio ursinus*) troops living on the desert floor of the Kuiseb Canyon in South West Africa. Stoning by these baboons is done from the rocky walls of the canyon where they sleep and retreat when they are threatened. Stones are lifted with one hand and dropped over the side. The stones tumble down the side of the cliff or fall directly to the canyon floor. The researchers recorded 23 such incidents involving the voluntary release of 124 stones.

A subadult male from a captive group of Guinea baboons (*Papio papio*) learned, by trial-and-error, to use a tool to rake in food. He then used the tool 104 times over 26 days, thereby providing the group with most of its food.

In April 2018, after four captive baboons managed to escape from their enclosure at Texas Biomedical Research Institute, a 55-gallon barrel left on its side in the pen as an enrichment device was found to have been stood erect next to the perimeter wall, enabling its use as a jumping platform to escape.

In the wild, mandrills have been observed to clean their ears with modified tools. Scientists filmed a large male mandrill at Chester Zoo stripping down a twig, apparently to make it narrower, and then using the modified stick to scrape dirt from underneath its toenails.

In Thailand and Myanmar, crab-eating macaques use stone tools to open nuts, oysters and other bivalves, and various types of sea snails (nerites, muricids, trochids, etc.) along the Andaman sea coast and offshore islands. A troop of wild macaques which regularly interact with humans have learnt to remove hairs from the human's heads, and use the hair to floss their teeth.

Elephants

Elephants show an ability to manufacture and use tools with their trunk and feet. Both wild and captive Asian elephants (*Elephas maximus*) use branches to swat flies or scratch themselves. Eight of 13 captive Asian elephants, maintained under a naturalistic environment, modified branches and switched with the altered branch, indicating this species is capable of the more rare behaviour of tool manufacture. There were different styles of modification of the branches, the most common of which was holding the main stem with the front foot and pulling off a side branch or distal end with the trunk. Elephants have been observed digging holes to drink water, then ripping bark from a tree, chewing it into the shape of a ball thereby manufacturing a "plug" to fill in the hole, and covering it with sand to avoid evaporation. They would later go back to the spot to drink.

Asian elephants may use tools in insightful problem solving. A captive male was observed moving a box to a position where it could be stood upon to reach food that had been deliberately hung out of reach.

Elephants have also been known to drop large rocks onto an electric fence to either ruin the fence or cut off the electricity.

Cetaceans

A community of Indo-Pacific bottlenose dolphins (*Tursiops* sp.) in Shark Bay, Western Australia, made up of approximately 41-54 animals, are known to use conical sponges (*Echinodictyum mesenterinum*) as tools while foraging. This behavior, termed “sponging”, occurs when a dolphin breaks off a sponge and wears it over its rostrum while foraging on the seafloor. Sponging behavior typically begins in the second year of life. During sponging, dolphins mainly target fish that lack swim bladders and burrow in the substrate. Therefore, the sponge may be used to protect their rostrums as they forage in a niche where echolocation and vision are less effective hunting techniques. Dolphins tend to carry the same sponge for multiple surfacings but sometimes change sponges. Spongers typically are more solitary, take deeper dives, and spend more time foraging than non-spongers. Despite these costs, spongers have similar calving success to non-spongers.

There is evidence that both ecological and cultural factors predict which dolphins use sponges as tools. Sponging occurs more frequently in areas with higher distribution of sponges, which tends to occur in deeper water channels. Sponging is heavily sex-biased to females. Genetic analyses suggest that all spongers are descendants of a single matriline, suggesting cultural transmission of the use of sponges as tools. Sponging may be socially learned from mother to offspring. Social grouping behavior suggests homophily (the tendency to associate with similar others) among dolphins that share socially learned skills such as sponge tool use. Sponging has only been observed in Shark Bay.

Indo-Pacific bottlenose dolphins in Shark Bay have also been observed carrying conch shells. In this behavior, dolphins insert their rostrum into the shell’s aperture. Although this behavior is rare, it appears to be used for foraging. Dolphins appear to use the conch shells to scoop fish from the substrate then carry the shell to retrieve the fish near the surface.

Sea Otters

A sea otter using a rock to break open a shell

Under each foreleg, the sea otter (*Enhydra lutris*) has a loose pouch of skin that extends across the chest. In this pouch (preferentially the left side), the animal stores collected food to bring to the surface. This pouch also holds a rock, unique to the otter, that is used to break open shellfish and clams. To open hard shells, it may pound its prey with both paws against the rock which it places on its chest. Furthermore, sea otters will use large stones to pry an abalone off its rock; they will hammer the abalone shell with observed rates of 45 blows in 15 seconds or 180 rpm, and do it in two or three dives. Releasing an abalone, which can cling to rock with a force equal to 4,000 times its own body weight, requires multiple dives by the otter.

Other Carnivores

Wild banded mongooses (*Mungos mungo*) regularly use anvils to open food items with a hard shell such as rhinoceros beetles, bird eggs, snail shells or pupating dung beetles. They use a range of anvils commonly including rocks and the stems of trees, but will also use the side-walls of gullys and even dried elephant dung. Pups as young as 2 months of age are already showing the behavioural patterns associated with using an anvil, however, successful smashing is usually shown in individuals older than 6 months of age.

Honey badgers both wild and captive have been filmed manipulating various objects to assist them in making climbs.

North American badgers (*Taxidea taxus*) hunt Richardson's ground squirrels (*Spermophilus richardsonii*). The most common hunting technique is excavation of burrow systems, but plugging of openings into ground-squirrel tunnels accounts for 5–23% of hunting actions. Badgers usually use soil from around the tunnel opening, or soil dragged 30–270 cm from a nearby mound to plug tunnels. The least common (6%), but most novel, form of plugging used by 1 badger involved movement of 37 objects from distances of 20–105 cm to plug openings into 23 ground-squirrel tunnels on 14 nights.

In 2011, researchers at the Dingo Discovery and Research Centre in Melbourne, Australia, filmed a dingo manipulating a table and using this to get food.

Moulting brown bears in Alaska have been observed using rocks to exfoliate.

In Birds

Tool use is found in at least thirty-three different families of birds. According to Jones and Kamil's definition, a bearded vulture dropping a bone on a rock would not be considered using a tool since the rock cannot be seen as an extension of the body. However, the use of a rock manipulated using the beak to crack an ostrich egg would qualify the Egyptian vulture as a tool user. Many other species, including parrots, corvids and a range of passerines, have been noted as tool users.

Bird Nests Show a Great Diversity in Complexity

Intricate nests of weaver birds

Relatively simple nest of a greater black-backed gull

Emperor penguins do not build a nest

Many birds (and other animals) build nests. It can be argued that this behaviour constitutes tool use according to the definitions given above; the birds "carry objects (twigs, leaves) for future use", the shape of the formed nest prevents the eggs from rolling away and thereby "extends the physical influence realized by the animal", and the twigs are bent and twisted to shape the nest, i.e. "modified to fit a purpose". The complexity of bird nests varies markedly, perhaps indicating a range in the sophistication of tool use. For example, compare the highly complex structures of weaver birds to the simple mats of herbaceous matter with a central cup constructed by gulls, and it is noteworthy that some birds do not build nests, e.g. emperor penguins. The classification of nests as tools has been disputed on the basis that the completed nest, or burrow, is not held or manipulated.

Woodpecker Finches

Perhaps the best known and most studied example of an avian tool user is the woodpecker finch (*Camarhynchus pallidus*) from the Galápagos Islands. If the bird uncovers prey in bark which is inaccessible, the bird then flies off to fetch a cactus spine which it may use in one of three different ways: as a goad to drive out an active insect (without necessarily touching it); as a spear with which to impale a slow-moving larva or similar animal; or as an implement with which to push, bring towards, nudge or otherwise manoeuvre an inactive insect from a crevice or hole. Tools that do not exactly fit the purpose are worked by the bird and adapted for the function, thus making the finch a "tool maker" as well as a "tool user". Some individuals have been observed to use a different type of tool with novel functional features such as barbed twigs from blackberry bushes, a plant that is not native to the islands. The twigs were first modified by removing side twigs and leaves and then used such that the barbs helped drag prey out of tree crevices.

There is a genetic predisposition for tool use in this species, which is then refined by individual trial-and-error learning during a sensitive phase early in development. This means that, rather than following a stereotypical behavioural pattern, tool use can be modified and adapted by learning.

The importance of tool use by woodpecker finch species differs between vegetation zones. In the arid zone, where food is limited and hard to access, tool use is essential, especially during the dry season. Up to half of the finches' prey is acquired with the help of tools, making them even more routine tool users than chimpanzees. The tools allow them to extract large, nutritious insect larvae from tree holes, making tool use more profitable than other foraging techniques. In contrast, in the humid zone, woodpecker finches rarely use tools, since food availability is high and prey is more easily obtainable. Here, the time and energy costs of tool use would be too high.

There have been reported cases of woodpecker finches brandishing a twig as a weapon.

Corvids

Corvids are a family of birds characterised by relatively large brains, remarkable behavioural plasticity (especially highly innovative foraging behaviour) and well-developed cognitive abilities.

New Caledonian Crow

New Caledonian crows (*Corvus moneduloides*) are perhaps the most studied corvid with respect to tool-use.

In the wild, they have been observed using sticks as tools to extract insects from tree bark. The birds poke the insects or larvae until they bite the stick in defence and can then be drawn out. This "larva fishing" is very similar to the "termite fishing" practised by chimpanzees. In the wild, they also manufacture tools from twigs, grass stems or similar plant structures, whereas captive individuals have been observed to use a variety of materials, including feathers and garden wire. Stick tools can either be non-hooked—being more or less straight and requiring only little modification—or hooked. Construction of the more complex hooked tools typically involves choosing a forked twig from which parts are removed and the remaining end is sculpted and sharpened. New Caledonian crows also use pandanus tools, made from barbed leaf edges of screw pines (*Pandanus* spp.) by precise ripping and cutting although the function of the pandanus tools is not understood.

While young birds in the wild normally learn to make stick tools from elders, a laboratory New Caledonian crow named "Betty" was filmed spontaneously improvising a hooked tool from a wire. It was known that this individual had no prior experience as she had been hand-reared. New Caledonian crows have been observed to use an easily available small tool to get a less easily available longer tool, and then use this to get an otherwise inaccessible longer tool to get food that was out of reach of the shorter tools. One bird, "Sam", spent 110 seconds inspecting the apparatus before completing each of the steps without any mistakes. This is an example of sequential tool use, which represents a higher cognitive function compared to many other forms of tool use and is the first time this has been observed in non-trained animals. Tool use has been observed in a non-foraging context, providing the first report of multi-context tool use in birds. Captive New Caledonian crows have used stick tools to make first contact with objects that were novel and hence potentially dangerous, while other individuals have been observed using a tool when food was within reach but placed next to a model snake. It has been claimed "Their [New Caledonian crow] tool-making skills exceed those of chimpanzees and are more similar to human tool manufacture than those of any other animal."

New Caledonian crows have also been observed performing tool use behaviour that had hitherto not been described in non-human animals. The behaviour is termed "insert-and-transport tool use". This involves the crow inserting a stick into an object and then walking or flying away holding both the tool and object on the tool.

Hawaiian Crow

Captive individuals of the critically endangered Hawaiian crow (*Corvus hawaiiensis*) use tools to extract food from holes drilled in logs. The juveniles exhibit tool use without training or social learning from adults. As 104 of the 109 surviving members of the species were tested, it is believed to be a species-wide ability.

Others

Other corvid species, such as rooks (*Corvus frugilegus*), can also make and use tools in the laboratory, showing a degree of sophistication similar to that of New Caledonian crows. While not confirmed to have used tools in the wild, captive blue jays (*Cyanocitta cristata*) have been observed using strips of newspaper as tools to obtain food.

Various corvids have reached for stones to place in a vessel of water so as to raise the surface level to drink from it or access a floating treat, enacting Aesop's Fable of The Crow and the Pitcher.

A wild American crow (*Corvus brachyrhynchos*) has been observed to modify and use a piece of wood as a probe. Green jays (*Cyanocorax yncas*) have been observed using sticks as tools to extract insects from tree bark. Large-billed crows in urban Japan have been filmed using an innovative technique to crack hard-shelled nuts by dropping them onto crosswalks (pedestrian crossings) and letting them be run over and cracked by cars. They then retrieve the cracked nuts when the cars are stopped at the red light. In some towns in America, crows drop walnuts onto busy streets so that the cars will crack the nuts. Hooded crows (*Corvus cornix*) use bait to catch fish. Individuals (who may have observed fish being fed bread by humans) will place the bread in the water to attract fish.

Common ravens (*Corvus corax*) are one of only a few species who make their own toys. They have been observed breaking off twigs to play with socially. A corvid has been filmed sliding repeatedly down a snow-covered roof while balancing on a lid or tray. Another incidence of play in birds has been filmed showing a corvid playing with a table tennis ball in partnership with a dog, a rare example of tool use for the purposes of play. Blue jays, like other corvids, are highly curious and are considered intelligent birds. Young blue jays playfully snatch brightly coloured or reflective objects, such as bottle caps or pieces of aluminium foil, and carry them around until they lose interest.

Warblers

A pouch containing a tailorbird's nest

The tailorbird (genus *Orthotomus*) takes a large growing leaf (or two or more small ones) and with its sharp bill pierces holes into opposite edges. It then grasps spider silk, silk from cocoons, or plant fibres with its bill, pulls this "thread" through the two holes, and knots it to prevent it from pulling through (although the use of knots is disputed). This process is repeated several times until the leaf or leaves forms a pouch or cup in which the bird then builds its nest. The leaves are sewn together in such a way that the upper surfaces are outwards making the structure difficult to see. The punctures made on the edge of the leaves are minute and do not cause browning of the leaves, further aiding camouflage. The processes used by the tailorbird have been classified as sewing, rivetting, lacing and matting. Once the stitch is made, the fibres fluff out on the outside and in effect they are more like rivets. Sometimes the fibres from one rivet are extended into an adjoining puncture and appear more like sewing. There are many variations in the nest and some may altogether lack the cradle of leaves. It is believed that only the female performs this sewing behaviour. The Latin binomial name of the common tailorbird, *Orthotomus sutorius*, means "straight-edged" "cobbler" rather than tailor. Some birds of the genus *Prinia* also practice this sewing and stitching behaviour.

Brown-headed Nuthatches

Brown-headed nuthatches (*Sitta pusilla*) have been observed to methodically use bark pieces to remove other flakes of bark from a tree. The birds insert the bark piece underneath an attached bark scale, using it like a wedge and lever, to expose hiding insects. Occasionally, they reuse the same piece of bark several times and sometimes even fly short distances carrying the bark flake in their beak. The evolutionary origin of this tool use might be related to these birds frequently wedging seeds into cracks in the bark to hammer them open with their beak, which can lead to bark coming off.

Brown-headed nuthatches have used a bark flake to conceal a seed cache.

Crested Lark

A Crested lark (*Galerida cristata*) has been photographed apparently holding in its bill a stone chip it was reportedly using to dislodge prey from paving joints.

Parrots

Kea, a highly inquisitive New Zealand mountain parrot, have been filmed stripping twigs and inserting them into gaps in box-like stoat traps to trigger them. Apparently, the kea's only reward is the banging sound of the trap being set off. In a similarly rare example of tool preparation, a captive Tanimbar corella (*Cacatua goffiniana*) was observed breaking off and "shaping" splinters of wood and small sticks to create rakes that were then used to retrieve otherwise unavailable food items on the other side of the aviary mesh. This behaviour has been filmed.

Many owners of household parrots have observed their pets using various tools to scratch various parts of their bodies. These tools include discarded feathers, bottle caps, popsicle sticks, matchsticks, cigarette packets and nuts in their shells.

Hyacinth macaws (*Anodorhynchus hyacinthinus*) have been repeatedly observed to use tools when breaking open nuts, for example, pieces of wood being used as a wedge. Several birds have wrapped a piece of leaf around a nut to hold it in place. This behaviour is also shown by palm cockatoos (*Probosciger aterrimus*). It seems that the hyacinth macaw has an innate tendency to use tools during manipulation of nuts, as naïve juveniles tried out a variety of objects in combination with nuts.

Egyptian Vultures

When an Egyptian vulture (*Neophron percnopterus*) encounters a large egg, it takes a stone into its beak and forcefully throws it at the egg until the shell is broken, usually taking a few minutes. This behaviour, first reported in 1966, seems to be largely innate and is displayed by naïve individuals. Its origin could be related to the throwing of eggs; rounded (egg-like) stones are preferred to jagged ones.

In a small population in Bulgaria, Egyptian vultures use twigs to collect sheep wool for padding their nests. Although both twigs and wool can serve as nesting material, this appears to be deliberate tool use. The birds approached bits of discarded wool with a twig in their beak, which was then either used as a rake, to gather the wool into heaps, or to roll up the wool. Wool was collected only after shearing or simulated shearing of sheep had taken place, but not after wool had simply been deposited in sheep enclosures.

Fire-foraging Raptors

In Australia the black kite (*Milvus migrans*), whistling kite (*Haliastur sphenurus*) and unrelated brown falcon (*Falco berigora*) are not only attracted to wildfires to source food, but will variously use their beaks or talons to carry burning sticks so as to spread fire, complicating human efforts to contain fires using firebreaks.

Owls

Burrowing owls (*Athene cunicularia*) frequently collect mammalian dung, which they use as a bait to attract dung beetles, a major item of prey.

Gulls

Seagulls have been known to drop live oyster shells on paved and hard surfaces so that cars can drive over them and break the shell. So many get dropped that it is difficult to drive safely near some waterways. Certain species (e.g. the herring gull) have exhibited tool use behavior, using pieces of bread as bait to catch goldfish.

Herons

The green heron (*Butorides virescens*) and its sister species the striated heron (*Butorides striata*) have been recorded using food (bread crusts), insects, leaves, and other small objects as bait to attract fish, which they then capture and eat.

In Reptiles

Tool use by American alligators and mugger crocodiles has been documented. During the breeding season, birds such as herons and egrets look for sticks to build their nests. Alligators and crocodiles collect sticks to use as bait to catch birds. The crocodilian positions itself near a rookery, partially submerges with the sticks balanced on its head, and when a bird approaches to take the stick, it springs its trap. This stick displaying strategy is the first known case of a predator not only using an object as a lure, but also taking into account the seasonal behavior of its prey.

In Fish

Several species of wrasses have been observed using rocks as anvils to crack bivalve (scallops, urchins and clams) shells. It was first filmed in an orange-dotted tuskfish (*Choerodon anchorago*) in 2009 by Giacomo Bernardi. The fish fans sand to unearth the bivalve, takes it into its mouth, swims several metres to a rock which it uses as an anvil by smashing the mollusc apart with sideward thrashes of the head. This behaviour has been recorded in a blackspot tuskfish (*Choerodon schoenleinii*) on Australia's Great Barrier Reef, yellowhead wrasse (*Halichoeres garnoti*) in Florida and a six-bar wrasse (*Thalassoma hardwicke*) in an aquarium setting. These species are at opposite ends of the phylogenetic tree in this family, so this behaviour may be a deep-seated trait in all wrasses.

It has been reported that freshwater stingrays use water as a tool by manipulating their bodies to direct a flow of water and extract food trapped amongst plants.

Prior to laying their eggs on a vertical rock face, male and female whitetail major damselfish clean the site by sand-blasting it. The fish pick up sand in their mouths and spit it against the rock face. Then they fan the area with their fins. Finally they remove the sand grains that remain stuck to the rock face by picking them off with their mouths.

Banded acara, (*Bujurquina vittata*), South American cichlids, lay their eggs on a loose leaf. The male and female of a mating pair often "test" leaves before spawning: they pull and lift and turn candidate leaves, possibly trying to select leaves that are easy to move. After spawning, both parents guard the eggs. When disturbed, the parent acara often seize one end of the egg-carrying leaf in their mouth and drag it to deeper and safer locations.

Archerfish are found in the tropical mangrove swamps of India and Australasia. They approach the surface, take aim at insects that sit on plants above the surface, squirt a jet of water at them, and grab them after the insects have been knocked off into the water. The jet of water is formed by the action of the tongue, which presses against a groove in the roof of the mouth. Some archerfish can hit insects up to 1.5 m above the water surface. They use more water, which gives more force to the impact, when aiming at larger prey. Triggerfish (*Pseudobalistes fuscus*) blow water to turn sea urchins over and expose their more vulnerable ventral side. Whether these later examples can be classified as tool use depends on which definition is being followed because there is no intermediate or manipulated object, however, they are examples of highly specialized natural adaptations.

In Invertebrates

Cephalopods

A small octopus (4-5 cm, c. 2-inch diameter) using a nut shell and clam shell as shelter

At least four veined octopus (*Amphioctopus marginatus*) individuals were witnessed retrieving coconut shells, manipulating them, stacking them, transporting them some distance (up to 20 metres), and then reassembling them to use as a shelter. The octopuses use coconut shells discarded by humans which have eventually settled in the ocean. They probe their arms down to loosen the mud, then rotate the shells out. After turning the shells so the open side faces upwards, the octopuses blow jets of mud out of the bowl before extending their arms around the shell—or if they have two halves, stacking them first, one inside the other. They then stiffen their legs and move away in a manner which has been called "stilt-walking". The octopuses eventually use the shells as a protective shelter in areas where little other shelter exists. If they just have one half, they simply turn it over and hide underneath. But if they are lucky enough to have retrieved two halves, they assemble them back into the original closed coconut form and sneak inside. This behaviour has been filmed. The authors of the research article claimed this behaviour falls under the definition of tool use because the shells are carried for later use. However, this argument remains contested by a number of other biologists who state that the shells actually provide continuous protection from abundant bottom-dwelling predators in their home range.

Octopuses deliberately place stones, shells and even bits of broken bottle to form a wall that constricts the aperture to the den, a type of tool use.

In laboratory studies, *Octopus mercatoris*, a small pygmy species of octopus, has been observed to block its lair using a plastic Lego brick.

Smaller individuals of the common blanket octopus (*Tremoctopus violaceus*) hold the tentacles of the Portuguese man o' war, to whose poison they are immune, both as protection and as a method of capturing prey.

Insects

Ants of the species *Conomyrma bicolor* pick up stones and other small objects with their mandibles and drop them down the vertical entrances of rival colonies, allowing workers to forage for food without competition.

Several species of ant are known to use substrate debris such as mud and leaves to transport water to their nest. A study in 2017 reported that when two species of Aphaenogaster ant are offered natural and artificial objects as tools for this activity, they choose items with a good soaking capacity. The ants develop a preference for artificial tools that cannot be found in their natural environment, indicating plasticity in their tool-use behaviour.

Hunting wasps of the genus *Prionyx* use weights (such as compacted sediment or a small pebble) to settle sand surrounding a recently provisioned burrow containing eggs and live prey in order to camouflage and seal the entrance. The wasp vibrates its wing muscles with an audible buzz while holding the weight in its mandibles, and applies the weight to the sand surrounding its burrow, causing the sand to vibrate and settle. Another hunting wasp, *Ammophila*, uses pebbles to close burrow entrances.

Insects can also learn to use tools. A study in 2017 showed that bumblebees of the species *Bombus terrestris* learned to move a small wooden ball to the goal in order to get sucrose reward.

References

- "New Bill In California to Protect Dogs in Police Altercations - PetGuide". PetGuide. 2017-02-22. Retrieved 2017-06-22
- Englar, Ryane E. (2017). Performing the Small Animal Physical Examination. John Wiley & Sons. ISBN 9781119295310
- Andrew right, A Criticism of the IASP's Definition of Pain, "Archived copy". Archived from the original on 22 August 2016. Retrieved 30 October 2017
- Matoba, Tomoyuki; Kutsukake, Nobuyuki; Hasegawa, Toshikazu (4 September 2013). "Head Rubbing and Licking Reinforce Social Bonds in a Group of Captive African Lions, Panthera leo". PLOS ONE. 8 (9): e73044. doi:10.1371/journal.pone.0073044. ISSN 1932-6203
- Bos, Ruud van den (1 March 1998). "The function of allogrooming in domestic cats (Felis silvestris catus); a study in a group of cats living in confinement". Journal of Ethology. 16 (1): 1–13. doi:10.1007/BF02896348. ISSN 0289-0771
- Merilaita, Sami; Vallin, Adrian; Kodandaramaiah, Ullasa; Dimitrova, Marina; Ruuskanen, Suvi; Laaksonen, Toni (26 July 2011). "Behavioral Ecology". Number of eyespots and their intimidating effect on naïve predators in the peacock butterfly. Oxford Journals. Retrieved 27 November 2011
- Leake, J. (14 March 2004). "Anglers to Face RSPCA Check". The Sunday Times. Archived from the original on 23 September 2015. Retrieved 15 September 2015
- Fendt, Markus; Kiyokawa, Yasushi; Endres, Thomas (2016). Scents that matter - from olfactory stimuli to genes, behaviors and beyond. Frontiers Media SA. p. 34. ISBN 9782889198139
- Krause, J.; Ruxton, G.; Rubenstein, D. (1998). "Is there always an influence of shoal size on predator hunting success?". Journal of Fish Biology. 52: 494–501. doi:10.1006/jfbi.1997.0595
- Crowell-Davis, Sharon L.; Curtis, Terry M.; Knowles, Rebecca J. (March 2004). "Social organization in the cat: a modern understanding" (PDF). Journal of Feline Medicine & Surgery. 6. doi:10.1016/j.jfms.2003.09.013. Retrieved 2 April 2017
- Martins, Marcio (1989). "Deimatic Behavior in Pleuroderma brachyops"(PDF). Journal of Herpetology. 23 (3): 305–307. doi:10.2307/1564457
- Snow, P.J.; Plenderleith, M.B.; Wright, L.L. (1993). "Quantitative study of primary sensory neurone populations of three species of elasmobranch fish". Journal of Comparative Neurology. 334: 97–103. doi:10.1002/cne.903340108
- Schaller, George B. (2009). The Serengeti Lion: A Study of Predator-Prey Relations. University of Chicago Press. ISBN 9780226736600. Retrieved 11 February 2018
- Gregory, Neville G. (2008). Physiology and Behaviour of Animal Suffering. John Wiley & Sons. p. 107. ISBN 9781405173025
- Ramachandran, V.S. (1 January 2009). "Self Awareness: The Last Frontier, Edge Foundation web essay". Retrieved 26 July 2011
- Kim, E.J.; Kim, E.S.; Covey, E.; Kim, J.J. (2010). "Social transmission of fear in rats: the role of 22-kHz ultrasonic distress vocalization". PLoS ONE. 5 (12): e15077. doi:10.1371/journal.pone.0015077. PMC 2995742 . PM ID 21152023
- Elwood, R.W. (2011). "Pain and suffering in invertebrates?" (PDF). Institute of Laboratory Animal Resources Journal. 52 (2): 175–84. Archived from the original (PDF) on 7 April 2012
- McGlone, John J. (1 April 1986). "Agonistic Behavior in Food Animals: Review of Research and Techniques". Journal of Animal Science. 62 (4): 1130–1139. doi:10.2527/jas1986.6241130x. ISSN 0021-8812

Permissions

All chapters in this book are published with permission under the Creative Commons Attribution Share Alike License or equivalent. Every chapter published in this book has been scrutinized by our experts. Their significance has been extensively debated. The topics covered herein carry significant information for a comprehensive understanding. They may even be implemented as practical applications or may be referred to as a beginning point for further studies.

We would like to thank the editorial team for lending their expertise to make the book truly unique. They have played a crucial role in the development of this book. Without their invaluable contributions this book wouldn't have been possible. They have made vital efforts to compile up to date information on the varied aspects of this subject to make this book a valuable addition to the collection of many professionals and students.

This book was conceptualized with the vision of imparting up-to-date and integrated information in this field. To ensure the same, a matchless editorial board was set up. Every individual on the board went through rigorous rounds of assessment to prove their worth. After which they invested a large part of their time researching and compiling the most relevant data for our readers.

The editorial board has been involved in producing this book since its inception. They have spent rigorous hours researching and exploring the diverse topics which have resulted in the successful publishing of this book. They have passed on their knowledge of decades through this book. To expedite this challenging task, the publisher supported the team at every step. A small team of assistant editors was also appointed to further simplify the editing procedure and attain best results for the readers.

Apart from the editorial board, the designing team has also invested a significant amount of their time in understanding the subject and creating the most relevant covers. They scrutinized every image to scout for the most suitable representation of the subject and create an appropriate cover for the book.

The publishing team has been an ardent support to the editorial, designing and production team. Their endless efforts to recruit the best for this project, has resulted in the accomplishment of this book. They are a veteran in the field of academics and their pool of knowledge is as vast as their experience in printing. Their expertise and guidance has proved useful at every step. Their uncompromising quality standards have made this book an exceptional effort. Their encouragement from time to time has been an inspiration for everyone.

The publisher and the editorial board hope that this book will prove to be a valuable piece of knowledge for students, practitioners and scholars across the globe.

Index